Springer Climate

Series Editor:

John Dodson, Menai, NSW, Australia

Springer Climate is an interdisciplinary book series dedicated to climate research. This includes climatology, climate change impacts, climate change management, climate change policy, regional climate, climate monitoring and modeling, palaeoclimatology etc. The series hosts high quality research monographs and edited volumes on Climate, and is crucial reading material for Researchers and students in the field, but also policy makers, and industries dealing with climatic issues. Springer Climate books are all peer-reviewed by specialists (see Editorial Advisory board). If you wish to submit a book project to this series, please contact your Publisher (elodie.tronche@springer.com).

More information about this series at http://www.springer.com/series/11741

Mozaharul Alam • Jeongho Lee • Puja Sawhney
Editors

Status of Climate Change Adaptation in Asia and the Pacific

Springer

Editors
Mozaharul Alam
United Nations Environment Programme
Bangkok, Thailand

Jeongho Lee
KACCC at Korean Environment Institute
Sejong, Republic of Korea

Puja Sawhney
Asia Pacific Adaptation Network (APAN)
Bangkok, Thailand

ISSN 2352-0698 ISSN 2352-0701 (electronic)
Springer Climate
ISBN 978-3-030-07592-7 ISBN 978-3-319-99347-8 (eBook)
https://doi.org/10.1007/978-3-319-99347-8

This Springer imprint is published by the registered company Springer Nature Switzerland AG
The registered company address is: Gewerbestrasse 11, 6330 Cham, Switzerland

Preface

Information and knowledge are integral parts of adaptation planning, monitoring progress, and communicating outcomes delivered through undertaking adaptation actions. Adaptation to climate change is often viewed as a knowledge-intensive exercise. Adaptation also entails engagement of actors ranging from global to local levels to create necessary means such as policies and governance, institutional and individual capacity, and access to finance and technologies to support adaptation actions. Environmental multilateralism has made significant progress by setting climate change as one of the top global issues. Several measures have been taken from global to local levels for assessing and understanding impacts, vulnerabilities, and adaptation to climate change since the Framework Convention on Climate Change and associated decisions came into force including the Marrakech Climate Accord in 2001 and the Cancun Framework for Adaptation in 2010. In 2015, the Paris Agreement established global adaptation goals including enhancing adaptive capacity, strengthening resilience, and reducing vulnerability to climate change. These indicators in the global adaptation goal, i.e., adaptive capacity, resilience, and vulnerability, present its inherent challenges for quantification and comparability, as well as aggregation of information both at national and global levels, to demonstrate the progress of adaptation.

Monitoring adaptation progress is a new and emerging issue, and needs are becoming more prominent due to the ratification of the Paris Agreement. Several initiatives in recent years have been working on this emerging need, and emphasis is placed on tracking adaptation finance and the flow of funds from developed countries to developing countries. Academic and research institutes have also developed different approaches for monitoring or tracking adaptation, focusing on outcomes, preparedness, and policy, but application of these approaches is limited. Additionally, a framework for assessing status of adaptation, and monitoring progress against predefined adaptation goals and/or objectives, and application in Asia and the Pacific region and subregional levels appears lacking. These gaps inspired the Asia Pacific Adaptation Network (APAN) and its partners to explore and make an attempt to address this topic. Strong determination and tireless efforts of APAN partners put together the rich contents in this book on "Status of Adaptation in Asia and the Pacific Region."

In the 2014 APAN, which unceasingly mobilized adaptation knowledge in collaboration with a wide range of partners, and the Korea Environment Institute (KEI) that has supported APAN since its inception in 2009 recognized the gaps and the need to develop a framework and apply it to assess the status of adaptation in Asia and the Pacific region. The APAN and its partners has also recognized the fact that Asia and the Pacific region has made good progress in several areas to advance adaptation measures including adaptation actions on the ground.

While social, economic, and environmental diversity is high among countries in Asia and the Pacific, vulnerabilities of social and economic sectors and the adaptive capacity of society and the economy vary across the regions, subregions, and countries. Efforts to deal with adverse impacts of climate change also vary, including adaptation objectives and goals. While expression of adaptation goals varies across subregions and countries, the overarching goals are meant to protect people, development, and ecology from adverse impacts of climate change. Furthermore, priorities for adaptation interventions vary across key economic and social development sectors and spatial contexts, while adaptation actions are strongly associated with the most vulnerable sectors and are linked to the economic development and livelihoods of large populations.

This book provides an in-depth assessment and overview of climate change adaptation in the Asia-Pacific region and the status of adaptation at subregional levels (Central Asia, Pacific, Northeast Asia, South Asia, and Southeast Asia); sectors, i.e., agriculture that supports livelihoods of many people; and large ecosystems, i.e., Hindu Kush Himalaya mountain ecosystem. In addition to an in-depth assessment of the status of adaptation, the authors have made efforts to identify future actions needed for enhancing adaptation measures and indicate the support available at international and regional levels. The book also aims to contribute to current and future discourse on assessing and communicating adaptation progress at different levels, without prejudging future outcomes, and implementing the Paris Agreement goals.

The editors express their heartfelt thanks to the Ministry of Environment, Government of Korea, for providing financial support, and researchers and staff members of the Korea Environment Institute (KEI) for organizing the authors' meeting and for the continued support provided to the authors. We acknowledge the dedication of all chapter authors and support provided by their respective institutes. These include Climate Action Network South Asia (CAN-SA); International Centre for Mountain Development (ICIMOD); Local Governments for Sustainability Southeast Asia Secretariat; Research Centre for Climate Change Adaptation, Keio University; School of Environmental Science and Management, University of Philippines Los Baños; Secretariat of the Pacific Regional Environment Programme (SPREP); Southeast Asian Regional Centre for Graduate Study and Research in Agriculture (SERCA); and the Regional Environmental Centre for Central Asia (CAREC). Thanks are due to the staff of Springer International Publishing for their timely efforts in publishing this book.

Bangkok, Thailand — Mozaharul Alam
Sejon-si, Republic of Korea — Jeongho Lee
Bangkok, Thailand — Puja Sawhney

Contents

Contributors

Ahsan Uddin Ahmed Centre for Global Change, Dhaka, Bangladesh

Mozaharul Alam United Nations Environment Programme, Bangkok, Thailand

Rico C. Ancog School of Environmental Science and Management, University of the Philippines Los Baños, Laguna, Philippines

Arivudai Nambi Appadurai Climate Resilience Practice, World Resources Institutes, Bengaluru, Karnataka, India

Ranell Martin M. Dedicatoria ICLEI–Local Governments for Sustainability Southeast Asia Secretariat (ICLEI SEAS), Quezon City, Philippines

Catherine B. Diomampo ICLEI–Local Governments for Sustainability Southeast Asia Secretariat (ICLEI SEAS), Quezon City, Philippines

William Galloway Research Center for Climate Change Adaptation (RCCCA), Keio University, Tokyo, Japan

Saleemul Huq International Centre for Climate Change and Development, Dhaka, Bangladesh

Abid Hussain International Centre for Integrated Mountain Development (ICIMOD), Kathmandu, Nepal

Ju Youn Kang Korea Adaptation Center for Climate Change (KACCC)/Korea Environment Institute (KEI), Sejong, Republic of Korea

Saniya Kartayeva The Regional Environmental Centre for Central Asia (CAREC), Almaty, Kazakhstan

Peniamina Dougalii Leavai Secretariat of the Pacific Regional Environment Programme, Apia, Samoa

Bidhubhusan Mahapatra International Centre for Integrated Mountain Development (ICIMOD), Kathmandu, Nepal

Nailya Mustaeva The Regional Environmental Centre for Central Asia (CAREC), Almaty, Kazakhstan

Sharmind Neelormi Economics Department, Jahangirnagar University, Dhaka, Bangladesh

Golam Rasul International Centre for Integrated Mountain Development (ICIMOD), Kathmandu, Nepal

Espen Ronneberg Secretariat of the Pacific Regional Environment Programme, Apia, Samoa

Clarissa D. Ruzol School of Environmental Science and Management, University of the Philippines Los Baños, Laguna, Philippines

Puja Sawhney Asia Pacific Adaptation Network (APAN), Bangkok, Thailand

Mariliza V. Ticsay Southeast Asian Regional Center for Graduate Study and Research in Agriculture (SEARCA), Manila, Philippines

Wanglin Yan Research Center for Climate Change Adaptation (RCCCA), Keio University, Tokyo, Japan

Chapter 1
Adaptation Journey

Mozaharul Alam

Abstract Continued advancement of scientific knowledge and understanding of climate change, visible sign of changes in physical system, adverse impacts on different sectors, and socio-economic development are advancing adaptation discussion both at international and national levels on a regular basis. This chapter on "adaptation journey" presents progress and advancement of adaptation to climate change discourse over time. This progression may present under four phases, i.e. (a) realization of climate change and adaptation (1979–2000); (b) beginning of adaptation actions and emergence of equal treatment (2001–2007); (c) enhancement of adaptation actions (2008–2015), and (d) global goal of adaptation, and monitoring and communicating adaptation progress (2015 onward). This chapter also highlights importance of mobilization of adaptation knowledge to support adaptation planning and implementation of actions.

Keywords Adaptation journey · Mobilization of knowledge · Adaptation finance · Adaptation technologies · Capacity building

1.1 Realization of Climate Change and Adaptation

It is difficult to determine a date or an event to be considered as inception of discussion on climate change, adverse impacts and vulnerability of climate change, and adaptation. However, recognition of changes in climatic system, potential impacts on development, and concern of humanity during 1970s and 1980s motivates multilateral organizations to take action for protecting climate within a global framework. Origination of the UN Framework Convention on Climate Change (UNFCCC) can be considered as an important step and a critical milestone at the global level to protect climate and deal with adverse impacts within wider contexts including sustainable development, biodiversity, and food production.

M. Alam (✉)
United Nations Environment Programme, Bangkok, Thailand
e-mail: alam31@un.org

M. Alam et al. (eds.), *Status of Climate Change Adaptation in Asia and the Pacific*, Springer Climate, https://doi.org/10.1007/978-3-319-99347-8_1

Climatology as an important branch of the science and practice of meteorology (Landsberg 1945), and the basic atmospheric physics of greenhouse warming known to the scientific communities for more than a century (Houghton 2009). The first world climate conference "a world conference of experts on climate and mankind" held in 1979 is considered as an important event that brought specialists from a wide range of disciplines. This event has collectively delivered an appeal to nations to take full advantage of existing knowledge, take steps for improvement of knowledge, and to prevent potential man-made changes in climate and adverse impacts of human well-being (WMO 1979). Another important event is the Villach Conference held in 1985 that focuses on role of carbon dioxide and other greenhouse gases in climate variations and associated impacts. Scientists attended the conference has produced a highly powerful report predicting temperature rises in the first half of the twenty-first century which would be greater than any in the human history (WMO 1986).

UN resolutions in 1987 on the Environmental Perspective to the Year 2000 and Beyond, and on the report of the World Commission on Environment and Development have played critical role in the UN General Assembly (UNGA) to recognize that climate change has an impact on development. In 1988, UNGA has further recognized that climate change is a common concern of humanity. It has further determined that necessary and timely action should be taken to deal with climate change within a global framework (UN 1988). Considering pervading influence of climate on human and many fields of human activities, it has established a process and set up an Intergovernmental Negotiating Committee (INC) to negotiate the UN Framework Convention on Climate Change (UNFCCC). Two years of hectic negotiations ended with an agreed text for the UN Framework Convention on Climate Change on 9 May 1992.

UN Framework Convention on Climate Change has identified several areas of actions to address climate change including reduction of greenhouse gases, which is the primary cause of human-induced climate change, and adaptation to manage impacts as well as take advantages of climate change. It has also agreed reporting country level actions to the UNFCCC secretariat through national communication. It is worth noting that the Villach conference suggested for a programme on climate change which to be promoted by governments and the scientific community in collaboration with the World Meteorological Organization (WMO), the United Nations Environment Programme (UNEP), and the International Council of Scientific Unions (ICSU).

Efforts of the global community after entry into force of the UNFCCC in 1994 were primarily focusing on implementation of Berlin Mandate. It agrees to begin a process which will enable country to taking actions for the period beyond 2000, including the adoption of a protocol or another legal instrument (UNFCCC 1995). IPCC Second Assessment Report (IPCC 1995) informed the policy makers that the atmospheric concentrations of greenhouse gases have grown significantly against 1992 base year. Report has also confirmed that this growth can be attributed largely to human activities, mostly fossil-fuel use, land-use change, and agriculture. The Berlin Mandate delivered the first legally binding protocol "Kyoto Protocol" in

1997 to reduce emission of greenhouse gas. A comprehensive implementation rules of the Protocol was adopted at COP7 in 2001 while the protocol came into force on 16 Feb 2005.

Action at national level during this period was primarily focusing on raising awareness among key policy and decision-makers towards ratification of the UN Framework Convention on Climate Change. Entry into force of the Framework Convention on Climate Change triggered the need to develop national greenhouse gas inventory as well as to assess impacts and adaptations to climate change. Meanwhile, Intergovernmental Panel on Climate Change (IPCC) released a Technical Guidelines for Assessing Climate Change Impacts and Adaptations that enable sectors, countries, and regions to estimate impacts and adaptations to climate change (Carter et al. 1994). This technical guideline is also known as IPCC seven steps methodology for assessing climate change impacts and adaptations. Several countries, for example, Bangladesh in Asia region, have applied this technical guideline for assessing impacts and adaptation options.

The U.S. Country Studies Programme (USCSP) was one of the largest programmes during this period that supported 49 countries for assessing climate change impacts. It has assessed impacts in one or more of eight sectors: coastal resources, agriculture, grasslands/livestock, water resources, forests, fisheries, wildlife, and health. These studies were focused on analysis of first-order biophysical effects, e.g. coastal inundation, crop yield, and runoff changes. Smith and Lazo carried out a review and synthesize the results of the impact assessments of the USCSP. The studies found that sea level rise could cause substantial inundation and erosion of valuable lands. It has also revealed that protecting developed areas would be economically sound. The studies showed mixed results for changes in crop yields, with a tendency towards decreased yields in African and Asian countries, particularly southern Asian countries, and mixed results in European and Latin American countries.

While it indicated that adaptation could significantly reduce impacts of yields, but it is not clear whether the adaptations are affordable or feasible. It has also revealed that future changes in runoff could result in increases in droughts or floods. It has also appeared that country has large capacity to deal with impacts of climate change on grasslands and livestock. Considering inadequate access to health care system in low-latitude countries, human health problems are likely to increase due to climate change. It also revealed that forests composition is likely to change including reduction of biomass as well as some wildlife species. While USCSP contributed in building capacity in developing countries to assess potential climate impacts, but many studies did not analyse the implications of biophysical impacts of climate change on socio-economic conditions, cross-sectoral integration of impacts, autonomous adaptation, or proactive adaptation. It has suggested that follow-on work should attempt to conduct more integrated studies of climate change impacts (Smith and Lazo 2001).

1.2 Marrakesh to Bali: Beginning of Adaptation Actions

Scientific assessment reports of the IPCC, Third Assessment Report in 2001, the Johannesburg World Summit on Sustainable Development in 2002, and the growing recognition of inadequate mitigation efforts by Parties to the UNFCCC pursued global community to take enhanced action on adaption. It has also recognized that the challenges of climate change would have to be addressed through a balance of mitigation and adaptation action. International awareness began to increase rapidly of the need for comprehensive climate information in support of national and international strategy for reducing greenhouse gas emissions and adapting to unavoidable climate change. Clear need also emerged on technology, finance, and capacity building.

"Marrakesh Accords", adopted in 2001, is a key milestone for adaptation to climate change. It has established Least Developed Country Fund (LDCF) to support LDCs for addressing their immediate and urgent adaptation needs including formulation and implementation of National Adaptation Programme of Action (NAPA). It has also established a Special Climate Change Fund (SCCF) for all developing countries to deal with climate change. In order to provide technical support to LDCs for formulation and implementation of NAPA, it has also established an Expert Group for LDCs.

Strengthening capacity for information generation and flow, and role of existing national institutions have also been recognized in Marrakesh Accord as important aspects to support adaptation actions. Role of traditional skills, knowledge and practices, and information sharing in developing countries have also been recognized as important elements to support adaptation action. It has also suggested that research centres and universities and other relevant organizations should play an important role in providing capacity-building services and facilitating the flow of knowledge, best practices, and information. Marrakesh Accord has also suggested strengthening the capacity of meteorological and hydrological services to collect, analyse, interpret, and disseminate weather and climate information to support implementation of national adaptation programmes of action.

It has also recognized needs to identify, assess, and priorities technologies both for mitigation and adaptation particularly for developing country Parties. Country driven-ness, involvement of different stakeholders, and consultative process have been suggested as important approach to identify the barriers to technology transfer and measures to address barriers. It has also recognized that barriers may be related to soft and hard technologies, such as mitigation and adaptation technologies, identify regulatory options, and develop fiscal and financial incentives and capacity building. It has also recognized the need to enhance skills for adoption, adaptation, installation, operation, and maintenance of specific environmentally sound technologies and a broadening of understanding of methodologies for evaluating alternative technological options.

To facilitate and catalyse the development and dissemination of climate information and knowledge that would inform and support adaptation policies and practices, Conference of the Parties established the Nairobi work programme (NWP) in December 2005, and named the "Nairobi work programme on impacts, vulnerability and adaptation to climate change" at COP12 in Nairobi in 2006. NWP provides knowledge support on impacts, vulnerability, and adaptation to climate change as a "knowledge-to-action hub" in collaboration with partners, Parties, and other experts.

Montreal Climate Change Conference adopted several decisions among which dialogue on long-term cooperative action to address climate change. A 5-year work programme has also been established on adaptation to climate change impacts were key adaptation-related decisions for enhancing implementation of the Convention and bring balance between mitigation and adaptation. An agreement on a 1-year process to develop operational modalities for the Adaptation Fund was another milestone. This Adaptation Fund was established in Marrakesh and is a unique fund that draws on proceeds generated by the clean development mechanism and supports concrete adaptation activities in developing countries.

IPCC assessment reports brought importance of climate change adaptation and impacts of climate change and its implications on development. Adaptation to climate change requires same level of attention as mitigation.

It also revealed that adaptation agenda was driven by Least Developed Countries (LDCs) during this period. Least Developed Countries taken steps for formulating National Adaptation Programme of Action (NAPA) with financial support Global Environment Facility (GEF). Several LDCs have formulated their NAPA and couple of adaptation projects hits the ground. In addition, several developing countries have taken steps to develop their climate change strategy and action plans, etc.

1.3 Bali to Paris: Enhanced Action

In the context of adaptation to climate change, significant progress has been made during this period to enhance support for adaptation planning, financing, technologies, and capacity building. In 2007, Bali Action Plan (BAP) has been adopted in Indonesia that brought both mitigation and adaptation almost on a same footage. It took 2 years of intense discussion and negotiations within the framework of long-term cooperative action to address climate change. Unequivocal conclusion of the IPCC Fourth Assessment Report about warming of the climate system as well as further delay in reducing emissions will significantly increase the risk of more severe climate change impacts provoked stronger decision to take at the highest political level of multilateral agreement.

Bali has also launched a process to reach an agreed outcome and adopt a decision in 2009. It aims to develop a shared vision for long-term cooperative action including a long-term global goal for emission reductions as without deep cuts in global emissions the ultimate objective of the Convention cannot be achieved. In the context of adaptation, it has identified and agreed several areas that would require international cooperation and support for enhancing adaptation actions. The area of cooperation and support may be categorized into three, i.e. (a) understand impacts and requirements for adaptation actions including technologies, finance, and capacity by undertaking vulnerability assessments, conducting exercise on prioritization of actions, assessment of financial needs; (b) develop response strategies including integration of adaptation actions into sectoral and national planning, and develop specific projects and programmes; (c) implementation of adaptation actions.

1.3.1 National Adaptation Plan

While Least Developed Countries (LDCs) were preparing National Adaptation Programme of Action (NAPA) to address their urgent and immediate adaptation needs, sharing early learning from formulation and implementation of adaptation projects, and evidences of climate change impacts presented by IPCC in its Fourth Assessment Report prompted discussion on need for medium- and long-term adaptation planning. Under the long-term cooperative action, "Cancun Adaptation Framework" has also been established with an aim to allow better planning and implementation of adaptation projects. It has broadened the scope of adaptation planning and implementation by agreeing that all countries face adaptation challenge, and international cooperation and support is urgently needed to implement adaptation actions. Though all countries need to adapt, it has recognized urgent and immediate needs of developing countries for reducing their vulnerability and building resilience. Considering special circumstances of Least Developed Country Parties, Cancun Adaptation Framework has further established a process to formulate and implement National Adaptation Plans (NAPs). It has further suggested that NAP should be building upon experience of LDCs in preparing and implementing National Adaptation Programmes of Action.

Work continued to elaborate purposes of NAP and approaches to apply for formulation. In 2011 in Durban, Parities to the UNFCCC agreed that the primary objectives of the national adaptation plan process will be twofold, i.e. (a) to reduce vulnerability to the impacts of climate change, by building adaptive capacity and resilience; and (b) to facilitate the integration of climate change adaptation into relevant policies, programmes, and activities. One of the key important aspects agreed by countries that adaptation planning is a continuous, progressive, and iterative process. While undertaking process of adaptation planning, each country will identify their priorities based on their country circumstances, national sustainable development objectives, plans, policies, and programmes. In 2012, LDC Expert Group (LEG) developed and published technical guideline for formulation of NAP.

1.3.2 Adaptation Financing

Operationalization of Adaptation Fund under Kyoto Protocol (KP), establishment of Green Climate Fund (GCF), and establishment of a mechanism to support meeting cost of adaptation under Paris Agreement are major adaptation financing instruments delivered during this period.

Though the Adaptation Fund was established in 2001 to finance concrete adaptation projects and programmes in developing countries but operationalization of the fund begun by establishment of the Adaptation Fund Board (AFB) in 2007 to supervise and manage the Adaptation Fund. Adoption of rules of procedure of the Adaptation Fund Board in 2008 laid down fundamental elements of institutional and operational structure to support concrete adaption interventions on the ground. Within 2 years, Adaptation Fund Board developed all necessary instruments for accreditation of international and national entities, and programming instruments for project appraisal and approval. In 2010, Adaptation Fund Board approved first four adaptation projects worth of about USD 14 million.

In 2010, Parties to the United Nations Framework Convention on Climate Change established Green Climate Fund (GCF) as an operating entity of the financial mechanism of the Convention under Article 11. Main purpose of the fund is to support projects, programmes, policies, and other activities in developing country Parties using thematic funding windows. This is in operation and aims to deliver equal amounts of funding to mitigation and adaptation, while being guided by the Convention's principles and provisions. The Green Climate Fund was given an important role in serving the Paris Agreement and supporting the goal of keeping increase global temperature well below 2 °C. Paris Agreement also requested Green Climate Fund to support formulation and implementation of National Adaptation Plans.

Paris Agreement has established a mechanism to contribute to the mitigation of greenhouse gas emissions and support sustainable development. This mechanism is another possible funding window on future to meet the costs of adaptation. Idea of accessing fund directly by country presented in this mechanism is similar to Adaptation Fund under Kyoto Protocol, where a share of the proceeds from mitigation activities is used to assist developing country Parties that are particularly vulnerable to the adverse effects of climate change to meet the costs of adaptation.

1.3.3 Technology Development and Transfer

Establishment of a Technology Mechanism in 2010 consisting of (1) a Technology Executive Committee, and (2) a Climate Technology Centre and Network was another milestone for technology development and transfer for both mitigation and adaptation. Parties under UNFCCC have decided that technology needs must be determined based on national circumstances and priorities. It has further suggested that research and development, demonstration, deployment, diffusion, and transfer of technology should be consistent with international obligations and to support both mitigation and adaptation actions.

In 2011, Parties to the Framework Convention on Climate Change reached to an agreement that arrangement to make the Technology Mechanism fully operational in 2012. It has also adopted the terms of reference for the CTCN and the selection process for the host of the Climate Technology Centre (CTC). In 2012, the UNFCCC Secretariat completed selection process of the host and awarded it to United Nations Environment Programme (UNEP) and United Nations Industrial Development Organization (UNIDO).

Due to limited focus on adaptation to climate change, definitions of technology and the concept of technologies for adaptation remain broad. In 2000, the IPCC in its special report on Methodological and Technological Issues in Technology Transfer, defines technology as "a piece of equipment, technique, practical knowledge or skills for performing a particular activity" (IPCC 2000). The UNFCCC has recognized the difficulty of defining the concept of adaptation technologies and proposes. It has defined adaptation technologies as "the application of technology to reduce the vulnerability, or enhance the resilience, of a natural or human system to the impacts of climate change" (UNFCCC 2005).

1.3.4 Capacity Building

Discussion on capacity building was initially focused on preparation of National Communication to the framework convention on climate change. Most of the capacity-building efforts were attached to preparation of national communication and technology needs assessment. Necessity of capacity building for assessing the costs of adaptation in developing countries is recognized as part of long-term cooperative action and agreed to enhance action on financial and technical support.

In Cancun, it has decided to enhance action on capacity-building support to developing countries for strengthening endogenous capacities at the sub-national, national, or regional levels. These actions will include strengthening (a) institutional capacity at various levels; (b) networks for the generation, sharing, and management of information and knowledge; (c) climate change communication, education, training, and public awareness; (d) existing and emerging capacity-building needs technology development and transfer as well as access to finance.

In 2015, Parties to the UNFCCC established the Paris Committee on Capacity Building. This committee will aim to address both current and emerging gaps and needs of developing countries. It has also decided to launch a work plan for the period 2016–2020. The work plan covers several areas including (a) an assessment on how to increase synergies through cooperation and avoid duplication; (b) identification of capacity gaps and needs as well as recommendation on way forward; (c) promote the development and dissemination of tools and methodologies for the implementation of capacity-building; (d) foster global, regional, national, and sub-national cooperation, (e) identification of opportunities to strengthen capacity at the national, regional, and sub-national levels.

In addition to several actions related to finance, technology, and capacity building, in 2010, Parties to the UNFCCC has established Adaptation Committee to promote the implementation of enhanced action on adaptation in a coherent manner. Discussion on loss and damage has also picked up during this time. A work programme established to identify approaches to address loss and damage associated with climate change impacts in developing countries. A Standing Committee on finance has also been established during this period to improve coherence and coordination in the delivery of climate change finance, rationalization of the financial mechanism, and mobilization of financial resources.

Development of climate change strategy and action plan continued during this period including formulation and enactment of climate change policies and acts. Efforts on integration of climate change adaptation into development planning and budgeting system has also emerged. Implementation of climate change adaptation projects has increased and learning by doing become a key approach to design and implement adaptation measures. Establishment of national adaptation fund or/and climate funds also has emerged during this period.

1.4 Paris and Beyond

1.4.1 Global Adaptation Goal

The Paris Agreement, ratified by 175 Parties[1] to the United Nations Framework Convention on Climate Change (UNFCCC), has established the global goal on adaptation of enhancing adaptive capacity, strengthening resilience, and reducing vulnerability to climate change. Three key indicators, i.e. adaptive capacity, resilience, and vulnerability of the global goal on adaptation come with their inherent challenges of quantification and comparability. To build mutual trust and confidence and to promote effective implementation, Parties have also agreed to an enhanced transparency framework for action, including adaptation, with built-in flexibility to accommodate specific circumstances and diversities of countries. While regular submission of information related to mitigation is mandatory (Article 13.7), adaptation reporting is not mandatory, as the Paris Agreement states that Parties "should", as appropriate, submit and update an "adaptation communication" (Article 7.10, 13.8). Parties has also requested the UNFCCC Secretariat to maintain a public registry of adaptation communications.

[1] http://unfccc.int/2860.php, accessed on 31 March 2018.

1.4.2 Adaptation Communication

Adaptation communication is one of the key building blocks of Paris Agreement which will allow countries to communicate their adaptation status and support need. It has decided to include priorities of a country, progress in implementation of adaptation measures, support received, and future support require in adaptation communication. Discussion is still going on and aiming to complete by 2018 focusing on defining the purpose of communication, different common elements which will applicable to all parties and additional elements those country may wish to include in their adaptation communication. Without prejudging the final agreed outcomes of the negotiations under Paris Agreement, discussions on purposes of adaptation communication are centred around (a) communicate national adaptation priorities, plans, actions, and implementation and support needs, (b) contribute and inform progress towards the global goal for adaptation, (c) provide input to the global stock-take, (d) recognition of adaptation efforts, (e) strengthen visibility and profile of adaptation/parity with mitigation, (f) enhance actions and support for developing countries, and (g) enhance learning and understanding of adaptation, sharing experiences and good practices.

1.5 Contents of the Book

Importance of mobilization of adaptation knowledge such as synthesize and develop new knowledge products to support adaptation planning and interventions has been recognized in several decisions of the United Nations Framework Convention of Climate Change (UNFCCC) including Nairobi Work Programme (NWP), and Lima Adaptation Knowledge Initiative (LAKI). It has also recognized that adaptation knowledge need to be mobilized for different actors at different levels and for different sectors to improve understanding and assessment of impacts, vulnerability and adaptation, and to support informed decision-making. Effective dissemination of climate information and knowledge products at regional, national, and sub-national levels on a timely manner is equally.

The Asia Pacific Adaptation Network (APAN), developed and launched by the United Nations Environment Programme (UN Environment) in 2009 under the Global Adaptation Network (GAN), is the first regional adaptation network. The Ministry of the Environment of Japan is supporting implementation of APAN's activities since its inception. Over the years, APAN has established a good partnership with key sub-regional organizations and become a key adaptation knowledge mobilizer in the Asia and the Pacific Region. It is responding to the request of CoP21 to the United Nations specialized organizations and agencies to support the adaptation efforts of Parties including strengthen cooperation on enhancing action on adaptation, sharing information, good practices, experiences and lessons learned related to science, planning, policies, and implementation. APAN is also supporting the Nairobi Work Programme (NWP) and Lima Adaptation Knowledge Initiative (LAKI) of the United Nations Framework Convention on Climate Change (UNFCCC) particularly development of knowledge products to improve understanding and assessment of impacts, vulnerability and adaptation, support informed decision-making as well as the effective dissemination of knowledge products at the regional, national, and sub-national levels. Asia Pacific Adaptation Forum become the largest gathering of adaptation practitioners for sharing and learning from each other.

This book is an outcome of collaborative efforts among Asia Pacific Adaptation Network (APAN), Korea Adaptation Centre for Climate Change (KACCC) of Korean Environment Institute (KEI), and APAN sub-regional and thematic partners. This book has tried to summarize various aspects of adaptation under eight essential areas of adjustment as argued in Chap. 2. This chapter on adaptation journey presents chronological development of adaptation trajectory mostly covering progress under the United Nations Framework Convention on Climate Change (UNFCCC) and efforts at national level. Chapter 2 presents an assessment framework, which has been applied for assessing status of climate change adaptation in five sub-regions, i.e. Central Asia, South Asia, Southeast Asia, Northeast Asia, and Pacific. This assessment framework has also been applied for assessing status of adaptation in Agriculture Sector in Asia, and Agriculture and Food Security in Hindu Kush Himalaya region. Chapter 3 provides an overview of status of adaptation in Asia and the Pacific region highlighting prevailing commonalities and differences. Five sub-regional level chapters, i.e. Chap. 4: Central Asia, Chap. 5: Northeast Asia, Chap. 6: Pacific Region, Chap. 7: South Asia, and Chap. 8: Southeast Asia, present status of adaptation at sub-regional level. Chapter 9 summarizes status of adaptation in Agriculture Sector while Chap. 10 provides status of adaptation in Hindu Kush Himalaya region. Chapter 11 provides an overview of country needs and mechanisms available at international and regional levels to support countries to enhance adaptation actions and address gaps.

References

Carter TR, Parry ML, Harasawa H, Nishioka S (1994) IPCC technical guidelines for assessing climate change impacts and adaptations. Department of Geography, University College London, London

Houghton JT (2009) Global warming: the complete briefing. Cambridge University Press, Cambridge

IPCC (1995) Climate change 1995. IPCC second assessment report

IPCC (2000) Special report on methodological and technological issues in technology transfer. A special report of IPCC working group III, published for the intergovernmental panel on climate change

Landsberg H (1945) Climatology. In: Berry FA, Bollay E, Beers NR (eds) Handbook of meteorology. McGraw Hill, London

Smith JB, Lazo JK (2001) A summary of climate change impact assessments from the U.S. country studies program. Clim Change 50:1–29 (Kluwer Academic)

UN (1988) UN resolution, A/RES/43/53 protection of global climate for present and future generations of mankind, New York
UNFCCC (1995) Report of the conference of the parties on its first session, Berlin, April 1995
UNFCCC (2005) Report on the seminar on the development and transfer of technologies for adaptation to climate change. FCCC/SBSTA/2005/8
WMO (1979) Declaration of world climate conference, IOC/SAB-IV/INF.3. World Meteorological Organization, Geneva
WMO (1986) Report of the international conference on the assessment of the role of carbon dioxide and of other greenhouse gases in climate variations and associated impacts. WMO-No. 661, Geneva

Chapter 2
Measuring Status of Climate Change Adaptation: An Assessment Framework

Mozaharul Alam and Saleemul Huq

Abstract The global adaptation goal is one of the key outcomes of the Paris Climate Change Agreement. This goal includes enhancement of adaptive capacity, strengthening of resilience, and reduction of vulnerability, while adaptation to climate change is defined as adjustment in natural or human systems in response to actual or expected climate stimuli or their effects. Tracking or measuring adaptation to climate change is a new aspect of this international discussion and ideas are emerging fast. Several initiatives are already in place and several relevant papers have been published in recent times. This assessment framework for measuring the status of climate change adaptation is an effort to contribute to the global discussion and understand different elements of adaptation. This assessment framework suggests a combination of readiness and outcome-based approaches to measure the status of adaptation by analyzing the adjustments required for achieving the adaptation objective(s). It also suggests how past adjustments can be used as evidence and demonstrate effectiveness in achieving similar adaptation objective(s). This assessment framework has been applied to analysis of the status of climate change adaptation in five subregions in Asia, and two sectors and elements are fine-tuned during its application.

Keywords Measuring climate change adaptation · Adjustment · Adaptation objectives · Status of adaptation

M. Alam (✉)
United Nations Environment Programme, Bangkok, Thailand
e-mail: alam31@un.org

S. Huq
International Centre for Climate Change and Development, Dhaka, Bangladesh
e-mail: saleemul.huq@iied.org

M. Alam et al. (eds.), *Status of Climate Change Adaptation in Asia and the Pacific*, Springer Climate, https://doi.org/10.1007/978-3-319-99347-8_2

2.1 Introduction

Avoidance of dangerous anthropogenic interference in the climatic system could have been the best method of lowering impacts associated with climate change and lessening the enormous adaptation needs arising today. More than 20 years ago, political leaders realized the need for addressing climate change and reached an agreement known as the United Nations Framework Convention on Climate Change (UNFCCC). The UNFCCC aims to stabilize concentrations of greenhouse gases in the atmosphere at a level that would prevent dangerous anthropogenic interference with the climate system. The main reasons behind the objective of stabilization were to (a) allow ecosystems to adapt naturally to climate change; (b) ensure that food production is not threatened; and (c) enable economic development to proceed in a sustainable manner (UNFCCC 1992). UNFCCC negotiations over more than two decades reached a political agreement to keep the global temperature rise below 2 °C compared to the preindustrial level and aspired to keep it within 1.5 °C (UNFCCC 2015). However, the progress toward stabilization of greenhouse gas concentrations that is required to keep the temperature rise below 2 °C is negligible. The latest scientific assessments have revealed that the temperature might even rise by more than 4 °C by the end of this century.

Since 1992, regular assessment reports published by the Intergovernmental Panel on Climate Change (IPCC), as well as other assessments including those from the International Energy Agency (IEA), have clearly revealed continuous growth of greenhouse gas emissions and concentrations at the global level. The IPCC's *Fifth Assessment Report* stated that carbon dioxide concentrations have increased by 40% since preindustrial times, primarily from fossil fuel emissions and secondarily from net land-use change emissions. The IPCC reports have also revealed that changes in climate have caused impacts on natural and human systems on all continents and across the oceans in recent decades. Evidence of climate change impacts is strongest and most comprehensive for natural systems and there are some impacts on human systems too. Continued emissions of greenhouse gases will cause further warming and changes in all components of the climate system (IPCC 2013, 2014).

The IPCC has defined adaptation as adjustment in natural or human systems in response to actual or expected climate stimuli or their effects. This adjustment aims to moderate harm or exploit beneficial opportunities. Various categories of adaptation include anticipatory and reactive adaptation, private and public adaptation, and autonomous and planned adaptation (Smith et al. 2001). There are several critical aspects in defining adaptation to climate change including adaptation for whom (natural and human systems), to what (temperature, rainfall, sea level rise, and salinity), at what scale (national, subnational, geospatial, etc.), and for which time frame. Actors' perception of climate change risks is an equally important factor for defining adaptation, as some risks may be routine and/or the consequences so minor that they are accepted, while risks that pose fundamental threats to actors' objectives or the sustainability of natural systems may be judged as intolerable (Klein et al. 2014).

Discussion on adaptation goal(s), adaptation objective(s), and adaptation gap(s) is appearing increasingly in the climate change discourse both in the scientific literature and in the climate change negotiations. The IPCC's *Fifth Assessment Report* stated that adaptation goals are often expressed in a framework of increasing resilience (Noble et al. 2014). It also appears that the difference between the current state and some predefined adaptation goals is a more likely way to define the adaptation gap. However, it is being noted that there is no common definition or set of elements used for defining adaptation goal(s), objective(s), and gap(s), and therefore all aspects stated above are important. While discussion is progressing on adaptation objective(s), several approaches are emerging for assessing the status of adaptation through tracking of climate finance; outcome-based approaches; preparedness-, process-, and policy-based approaches; and developmental outputs and outcomes of climate adaptation investment approaches, etc. (CFU 2013, Ford et al. 2013, Brooks and Fisher 2014).

2.2 Typologies for Tracking Adaptation

Though tracking adaptation is new and emerging, there are several relevant initiatives and papers that have been published in recent times. These include information on tracking climate finance by Climate Funds Update (CFU); outcome-based approaches; preparedness-, process-, and policy-based approaches (Ford et al. 2013); and developmental outputs and outcomes of a climate adaptation investment approach called Tracking Adaptation and Measuring Development (TAMD) (Brooks et al. 2013).

Climate Funds Update (CFU) is an independent website that provides information on the growing number of international climate finance initiatives designed to help developing countries to address the challenges of climate change. It is a joint initiative of the Heinrich Böll Stiftung (HBF) and the Overseas Development Institute (ODI), in an effort to increase the transparency of climate finance flows. It provides information on funds from the stage of donors' pledges for funding through to the actual disbursement of financing for projects.

The Tracking Adaptation and Measuring Development (TAMD) framework offers a "twin track" approach that focuses on assessing and comparing the effectiveness of interventions that directly or indirectly help populations adapt to climate change. It differs from other assessment frameworks by emphasizing the need to assess development interventions in the light of changing climate risks. This is to avoid missing effective interventions whose outcomes are obscured by increasing risks and vulnerability (Brooks et al. 2013).

Ford et al. (2013) have developed a typology of approaches by which climate change adaptation can be tracked globally at a national level. They have come up with two main typologies: (a) outcome-based approaches, which directly measure adaptation progress and effectiveness of avoidance of climate change impacts; and (b) preparedness-, process-, and policy-based approaches, which systematically measure adaptation readiness, processes undertaken to advance adaptation, policies and programs implemented for adaptation, and measures of the impacts of these policies and programs on changing vulnerability. Table 2.1 lists the key elements/indicators of the tracking approaches mentioned above.

Table 2.1 Key elements/indicators of several tracking approaches for adaptation to climate change

Name of framework	Key elements/indicators for tracking adaptation
CFU	• Where and by whom climate change funds are being developed • Scale of proposed and actual financing • What the funds support, in terms of focus, regions, and projects
TAMD	• Tracking elements at the national level: climate change integration (mainstreaming) into planning; budgeting and finance; institutional knowledge/capacity; use of climate information; planning under uncertainty (using appropriate information and methodologies); participation (of relevant stakeholders in national planning processes); awareness among stakeholders
Outcome evaluation: reduced negative climate change impacts	• Climate-related losses, mortality, and morbidity, over time and in relation to adaptation • Impacts of climatic hazard events before and after adaptation
Adaptation readiness: presence of key governance factors essential for effective and successful adaptation	• Evidence of political leadership; institutional organization; stakeholder involvement; climate change information; appropriate use of decision-making techniques; and consideration of barriers to adaptation, funding, technology development, and adaptation research
Process-based approaches: process through which adaptations are developed and implemented in pursuance of a desired outcome or objective	• Comparison of adaptation characteristics and steps of development to theoretically and empirically derive characteristics of adaptation success and best practice
Analyzing policies and programmatic approaches: monitoring and comparison of reported adaptation actions and their characteristics	• Analysis of characteristics of reported adaptations and comparison across regions, by vulnerability categories, over time, and with respect to adaptation "obligations"
Examining measures of changing vulnerability: measurement of change in vulnerability in relation to adaptation	• Monitoring of aggregate vulnerability indexes in relation to adaptation actions • Focus on specific indicators that capture generic determinants of vulnerability (e.g., limited access to education, poverty, health, and inequality) • Examination of specific components of sensitivity and capacity for adaptation to climate change impacts

CFU Climate Funds Update, *TAMD* Tracking Adaptation and Measuring Development

2.3 Elements of the Assessment Framework

This assessment framework has suggested four key elements for assessing the status of adaptation: (a) adaptation objective(s); (b) adjustments required for achievement; (c) current trends in adjustment, adequacy, and effectiveness in addressing similar adaptation objective(s); and (d) adequacy of current trends of adjustment to address future adaptation objective(s). It focuses on a readiness approach that systematically assesses adjustments undertaken to advance adaptations in policies and program implementation, results of adjustments, and use of past trends in adjustments and results as proxy indicators for adequacy and efficacy of readiness in achieving similar adaptation objective(s) for climate change in the future. Figure 2.1 visualizes the key elements of this assessment framework.

2.3.1 Assessment Indicators

The IPCC's *Fifth Assessment Report* identified a set of factors that enhance the potential for an actor to plan and implement actions to achieve their adaptation objective(s), as well as a set of factors that make it harder to plan and implement adaptation actions. Factors that enhance the potential are (1) awareness—communication, education, and awareness raising; (2) capacity—human and institutional capacity building, including preparedness, resource provision, and development of human and social capital; (3) tools—decision making, vulnerability and risk analysis, decision support, and early warning tools; (4) policy—integration and mainstreaming of policy, governance, and planning processes including

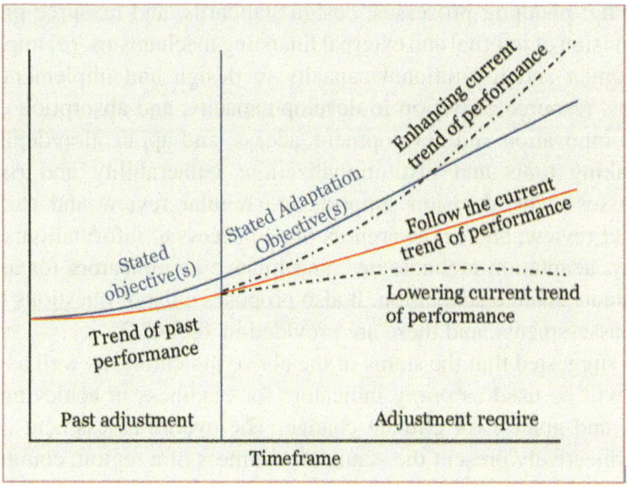

Fig. 2.1 Schematic presentation of the assessment framework and its elements

sustainable development, resource and infrastructure planning, and design standards; (5) learning—mutual experiential learning and management of knowledge of climate vulnerability, adaptation options, disaster risk response, and monitoring and evaluation; and (6) innovation—development and dissemination of new information, technology development, and technology application. It also stated that an opportunity is distinct from an adaptation option, which is a specific means of achieving an adaptation objective, such as an early warning system as a means of reducing vulnerability to tropical cyclones (Klein et al. 2014).

The list of constraints that hinder implementation of an adaptation option include (1) economic—existing livelihoods, economic structures, and economic mobility; (2) social and cultural—social norms, identity, place attachment, beliefs, world views, values, awareness, education, social justice, and social support; (3) human capacity—individual, organizational, and societal capabilities to set and achieve adaptation objectives over time, including training, education, and skill development; (4) governance, institutions, and policy—existing laws, regulations, procedural requirements, governance scope, effectiveness, institutional arrangements, adaptive capacity, and absorption capacity; (5) financial—lack of financial resources; (6) information/awareness/technology—lack of awareness of or access to information or technology; (7) physical—presence of physical barriers; and (8) biological—temperature, precipitation, salinity, acidity, and intensity and frequency of extreme events including storms, drought, and wind (Klein et al. 2014). The factors listed above that either enhance the potential for actors to plan and implement or act as constraints are not mutually exclusive; rather, many of them are two sides of the same coin or, more specifically, the presence or absence of these factors.

This assessment framework suggests use of (a) policy, strategy, legal provision, and governance: existing policy, strategy, laws, regulations, procedural requirements, and governance scope; (b) institutional arrangements: existing climate coordination, monitoring and evaluation, and feedback mechanisms; (c) integration and mainstreaming: planning processes, design standards, and resource provision; (d) finance: provision of internal and external financing mechanisms; (e) implementation capacity: human and institutional capacity to design and implement adaptation interventions, resource provision to develop capacity, and absorption capacity; (f) technology: innovation and development, access, and application/deployment; (g) decision-making tools and institutionalization: vulnerability and risk analysis, economic assessment, decision support, and regular review and early warning/climate under review; and (h) awareness of and access to information: impacts and vulnerability, adaptation technologies, and finance as indicators for assessing the status of climate change adaptation. It also proposes a list of questions to use while conducting assessments, and these are provided in Table 2.2.

It is also suggested that the status of the above indicators, as well as their trends of change, will be used as proxy indicators for readiness in achieving adaptation objective(s) and goal(s) for climate change. The overall assessment in the above areas will collectively present the status of readiness of a region, country, or sector in achieving their set adaptation objective(s).

Table 2.2 Indicators, subindicators, and questions to use for assessing the status of adjustments and trends

Indicators	Subindicators	Questions
(a) Policy, strategy, legal provision, and governance	1. Existing policy and strategy 2. Laws and regulations 3. Procedural requirements 4. Governance scope	1. Are there policy and strategy to deal with climate change at national, subnational, or local levels? 2. Is there a legal provision (laws and regulations) that binds implementation of policy and strategy? 3. Is there a procedure (such as climate screening of projects and an approval system like the EIA and approval of development projects) at any level? 4. Is there an oversight body to ensure compliance with policy, legal provisions, and procedural requirements, as well as dealing with noncompliance aspects? 5. Are the existing policy, strategy, legal provision, and governance adequate?
(b) Institutional arrangements	1. Existing climate coordination mechanism 2. Monitoring and evaluation, and feedback mechanism	1. Is there a body mandated to coordinate climate change to ensure synergy among interventions at national, subnational, or local levels, as well as between and among sectors? 2. Does the coordinating body have adequate convening power (authority/hierarchical importance) over key departments or ministries? 3. Does the coordinating body include multiple sectors? 4. Is there a mechanism to monitor progress, evaluate results/outcomes/outputs, and provide feedback for improvement? 5. Is the current arrangement functioning well and adequate?
(c) Integration and mainstreaming	1. Planning processes 2. Design standards 3. Resource provision	1. Has climate change been integrated into national and sectoral planning processes? 2. Does the planning process follow a set of design standards (with respect to temperature rise, sea level change, frequency of extreme events, etc.) depending on the types and location of projects? 3. Is there provision of resources for integrating climate change during the planning process? 4. Is current practice adequate?

(continued)

Table 2.2 (continued)

Indicators	Subindicators	Questions
(d) Finance	1. Provision of internal finance 2. Provision of external finance 3. Development finance for adaptation	1. Is internal[a] funding available to implement climate change adaptation interventions? 2. Is external funding available to implement climate change adaptation interventions? 3. Does external funding play a significant role in climate change adaptation activities? 4. Is there a fund-raising strategy in place to meet additional funding requirements? 5. Is development financing contributing to adaptation?
(e) Implementation capacity	1. Human and institutional capacity for design, implementation, and monitoring and evaluation of adaptation interventions 2. Resource provision[b] to develop capacity 3. Absorption capacity	1. Do sectoral ministries and departments (key vulnerable sectors) have institutional and human capacity to design and implement adaptation projects? 2. Is there provision of resources to build/enhance institutional and human capacity? 3. Does the current absorption capacity demonstrate adequacy?
(f) Technology	1. Innovation and development 2. Access and application/ deployment	1. Do the country/sectors have technology innovation facility and capacity? 2. Does the country have a financial investment provision and adequacy for technology development? 3. Do the country/sectors have provision of access to technology?[c] 4. Have the country/sectors accessed technology/technologies and applied them successfully?
(g) Decision-making tools and institutionalization	1. Vulnerability and risk analysis 2. Economic assessment 3. Decision support 4. Climate trends under review	1. Do the country/sectors undertake vulnerability and risk assessment regularly?[d] 2. Does the assessment include economic aspects to support decision making? 3. Is there any system to carry out regular review and flag emerging issues? 4. Have the above been institutionalized?

(continued)

Table 2.2 (continued)

Indicators	Subindicators	Questions
(h) Awareness of, knowledge of, and access to information	1. Impacts and vulnerability 2. Adaptation technologies 3. Source of adaptation finance 4. Sources of knowledge and means of dissemination	1. Are stakeholders[e] aware of climate change and its potential impacts and implications? 2. Are the stakeholders aware of available and potential adaptation technologies and access? 3. Are the stakeholders aware of available sources of funding (internal and external)? 4. Are mechanisms and systems in place for knowledge development and dissemination?

EIA Environmental Impact Assessment
[a]Internal funding includes governmental funds and other resources
[b]Resource provision refers to internal financial resource provision to retain/maintain implementation capacity
[c]Technology here means technology from outside
[d]This refers to conducting assessments beyond national communication
[e]Stakeholders includes policy makers, practitioners, and private sectors on the ground

2.3.2 *Assessment Steps*

To assess the status of climate change adaptation and apply the overall framework, the following five steps are suggested and examples are provided with possible sources of information to use.

2.3.2.1 Step I: Assessment of Adaptation Objective(s)

Assessment of adaptation objective(s) is the first step. Adaptation objectives can be assessed at different levels (national, subnational, and local) and for different sectors. Three broader objectives can be applied to keep adaptation objectives aligned with the overall goal of the convention but defined in the context of the adaptation. These adaptation objectives can be related to (a) the ecosystem—enhancing the resilience of ecosystems to maintain ecosystem services for dependent natural and human systems and human-assisted support for natural systems to adapt naturally to climate change; (b) food production—maintaining production of food and/or ensuring availability of food; and (c) maintaining economic development in a sustainable manner by addressing climate change impacts in key economic sectors. Table 2.3 provides an example, using the impacts of climate change on food production.

Table 2.3 Examples of adaptation objectives for food production and food systems

Sector	Examples of adaptation objectives	Sources of information
Food production and food systems	Maintain or enhance crop yields to sustain current level of food supply from domestic sources	• Climate change policy, strategy, and action plan • National development plan • Sectoral policy, strategy, and plan • National communication to UNFCCC • NAPA
	Reduce climate risk and vulnerability of key crop-producing agroecological zones	
	Provide a financial safety net and alternative job creation for farmers who will be unable to continue farming	

NAPA National Adaptation Program of Action, *UNFCCC* United Nations Framework Convention on Climate Change

Table 2.4 Examples of adaptation objectives and adaptation options for food production and food systems

Sector	Adaptation objective	Adaptation options	Sources of information
Food production and food systems	Maintain or enhance crop yields to sustain current level of food supply from domestic sources	Introduction of biotechnology and genetically modified crops, high-yielding varieties, climate stress resistance varieties, irrigation systems, etc.	• National communication to UNFCCC • NAPA • Technology Needs Assessment Report
	Reduce climate risk and vulnerability of key crop-producing agroecological zones	Introduction of climate stress resistance varieties, irrigation systems, biotechnology, genetically modified crops, etc.	
	Provide a financial safety net and alternative job creation for farmers who will be unable to continue farming	Subsidized drought assistance; crop insurance Increased use of chemical fertilizers and pesticides	

NAPA National Adaptation Program of Action, *UNFCCC* United Nations Framework Convention on Climate Change

2.3.2.2 Step II: Assessment of Options to Achieve Adaptation Objectives

The second step of the framework is identification and assessment of adaptation option(s) for achieving defined adaptation objective(s). Table 2.4 provides examples of adaptation objectives, options, and possible sources of information to use while undertaking assessment under step II.

Table 2.5 Example of adjustment require for implementing adaptation options for maintaining or enhancing crop yields as an adaptation objective

Adaptation objective: Maintain or enhance crop yields to sustain the current level of food supply from domestic sources			
Adaptation options	Required adjustment	Use the questions listed in Table 2.2 (yes/ partially/no)	If "yes" what are the status and types of adjustment required? If "no" how is the country conducting its VIA, and how can this be institutionalized?
Introduction of biotechnology, genetically modified crops, high-yielding varieties, climate stress resistance varieties	(a) Policy, strategy, legal provision, and governance	Yes	The country has developed a climate change policy, strategy, and action plan including good legislative provisions
	(b) Institutional arrangements		
	(c) Integration and mainstreaming	Partially	The country has integrated climate change into the national development policies and plans but has not integrated it into local-level development policies and plans
	(d) Finance		
	(e) Implementation capacity		
	(f) Technology		
	(g) Decision-making tools and institutionalization	No	Example: The country carries out vulnerability and risk assessment including key sectors regularly as a part of its national communication. However, the assessment does not include economic aspects to support decision making. There are existing institutes and the capacity for conducting VIA but this is done only when there is a specific project. Moreover, there is no system in place to carry out regular review and flag emerging issues
	(h) Awareness of and access to information		

VIA Vulnerability and Impact Assessment

2.3.2.3 Step III: Identification of Adjustments Required in Materializing Options for Achieving Adaptation Objectives

The third step of the assessment is to identify adjustments required for implementing the adaptation options identified to achieve the adaptation objective(s). Table 2.5 provides a list of areas that would require adjustment for materializing options to implement.

2.3.2.4 Step IV: Assessment of Past Trends of Adjustment

A key question is how an actor or group of actors have maintained similar adaptation objective(s) over the last 10 years. Examples are maintenance of food production or introduction of high-yielding varieties and/or climate stress resistance varieties. One of the reasons behind proposing 10 years is that the adaptation journey progresses together with other social and economic development efforts of a country. What were the key elements that enabled the actor(s) to maintain the set goal/objectives and/or introduction of option(s)? Table 2.6 provides an example.

Table 2.6 Example of adjustment require for implementing adaptation option for maintaining or enhacing crop yields as an adaptation objective

Adaptation objective: Maintain or enhance crop yields to sustain the current level of food supply from domestic sources		
Adaptation option	Trend of adjustment	Describe what has been done and use questions similar to those presented in Table 2.2
Introduction of high-yielding varieties	(a) Policy, strategy, legal provision, and governance	The country has developed a policy for introduction of new varieties to address drought
	(b) Institutional arrangements	
	(c) Integration and mainstreaming	
	(d) Finance	The country has received *x* amount of financial support from *y* development partners for *z* years for introducing new crop varieties
	(e) Implementation capacity	
	(f) Technology	
	(g) Decision-making tools and institutionalization	
	(h) Awareness of and access to information	
Overall results		

2.3.2.5 Step V: Assessment of Current Capacity for Addressing the Future

This is the most challenging part of the assessment, as this will require good understanding of the level of impact, options available for achieving adaptation objectives, and assessment of (a) whether the current trend will continue; (b) the adequacy of the current trend of adjustment in achieving the adaptation objective(s); and (c) what the key turning points are that can expedite the adjustment or become a setback.

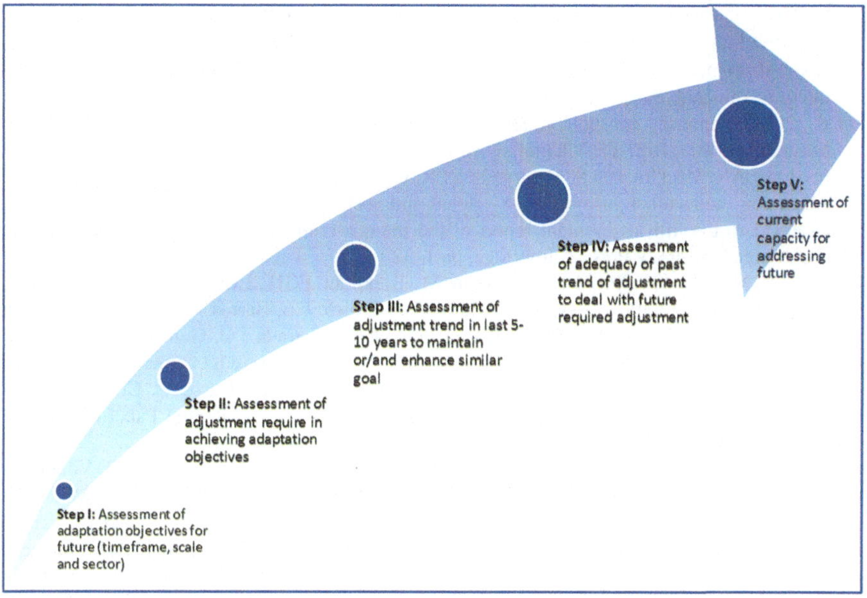

2.4 Conclusion

This assessment framework for measuring adaptation status is developed with an aim to apply it for assessing the adaptation status of five subregions in Asia and the Pacific Region, as well as two sectors and one thematic area. This assessment framework has been fine-tuned throughout the assessment and by bringing learning from application at regional levels as well as for sectors (agriculture and water) and themes (mountain agriculture and food security).

References

Brooks N, Anderson S, Burton I, Fisher S, Rai N, Tellam I (2013) An operational framework for Tracking Adaptation and Measuring Development (TAMD). IIED Climate Change Working Paper No. 5. IIED, London

Brooks N, Fisher S (2014) Tracking Adaptation and Measuring Development: a step-by-step guide. IIED, London

CFU 2013, https://climatefundsupdate.org/

Ford JD, Berrang-Ford L, Lesnikowski A, Barrera M, Heymann SJ (2013) How to track adaptation to climate change: a typology of approaches for national-level application. Ecol Soc 18(3):40 https://doi.org/10.5751/ES-05732-180340

IPCC (2013) Summary for policymakers. In: Stocker TF, Qin D, Plattner G-K, Tignor M, Allen SK, Boschung J, Nauels A, Xia Y, Bex V, Midgley PM (eds) Climate change 2013: the physical science basis. Contribution of Working Group I to the fifth assessment report of the Intergovernmental Panel on Climate Change. Cambridge University Press, Cambridge

IPCC (2014) Summary for policymakers. In: Field CB, Barros VR, Dokken DJ, Mach KJ, Mastrandrea MD, Bilir TE, Chatterjee M, Ebi KL, Estrada YO, Genova RC, Girma B, Kissel ES, Levy AN, MacCracken S, Mastrandrea PR, White LL (eds) Climate change 2014: impacts, adaptation, and vulnerability. Part A: global and sectoral aspects. Contribution of Working Group II to the fifth assessment report of the Intergovernmental Panel on Climate Change. Cambridge University Press, Cambridge, pp 1–32

Klein RJT, Midgley GF, Preston BL, Alam M, Berkhout FGH, Dow K, Shaw MR (2014) Adaptation opportunities, constraints, and limits. In: Field CB, Barros VR, Dokken DJ, Mach KJ, Mastrandrea MD, Bilir TE, Chatterjee M, Ebi KL, Estrada YO, Genova RC, Girma B, Kissel ES, Levy AN, MacCracken S, Mastrandrea PR, White LL (eds) Climate change 2014: impacts, adaptation, and vulnerability. Part A: global and sectoral aspects. Contribution of Working Group II to the fifth assessment report of the Intergovernmental Panel on Climate Change. Cambridge University Press, Cambridge, pp 899–943

Noble IR, Huq S, Anokhin YA, Carmin J, Goudou D, Lansigan FP, Osman-Elasha B, Villamizar A (2014) Adaptation needs and options. In: Field CB, Barros VR, Dokken DJ, Mach KJ, Mastrandrea MD, Bilir TE, Chatterjee M, Ebi KL, Estrada YO, Genova RC, Girma B, Kissel ES, Levy AN, MacCracken S, Mastrandrea PR, White LL (eds) Climate change 2014: impacts, adaptation, and vulnerability. Part A: global and sectoral aspects. Contribution of Working Group II to the fifth assessment report of the Intergovernmental Panel on Climate Change. Cambridge University Press, Cambridge, pp 833–868

Smith B, Pilifosova O, Burton I, Challenger B, Huq S, Klein RJT, Yohe G (2001) Adaptation to climate change in the context of sustainable development and equity. In: Climate change 2001: impacts, adaptation, and vulnerability. Contribution of Working Group II to the third assessment report of the Intergovernmental Panel on Climate Change. Cambridge University Press, Cambridge

UNFCCC (1992) The United Nations Framework Convention on Climate Change. UNFCCC Secretariat, Bonn

UNFCCC (2015) FCCC/CP/2015/L.9/Rev.1, Bonn, Germany

Chapter 3
Regional Overview

Mozaharul Alam and Puja Sawhney

Abstract Asia and the Pacific region is very diverse in many contexts—social, economic, and environment. Similarly impacts of climate change, vulnerabilities of social and economic sectors, and adaptive capacity of society and economy vary across sub-regions and countries within and among sub-regions. While scale of impacts, vulnerabilities, and adaptive capacity vary by sub-region, country, sector, and community, similarities are found among adaptation measures to deal with climate change impacts and vulnerabilities. This chapter summarizes prevailing commonalities and differences among sub-regions and countries in Asia and the Pacific region. These include similarities and differences in impacts, vulnerabilities, adaptation goals and priorities, readiness to response as well as gaps based on sub-regional and sectoral assessments of status of adaptation.

Keywords Climate change · Vulnerability · Adaptation · Adaptation goal · Policy · Technologies

3.1 Introduction

Asia and the Pacific region is very diverse in many contexts. Socio-economic condition, key economic sectors and growth, environmental situation varied widely among countries in five sub-regions, i.e. Central Asia, Northeast Asia, South Asia, Southeast Asia, and the Pacific. Except Australia, Japan, Republic of Korea, and New Zealand, rest falls under developing country category including Least Developed Countries. Several developing countries in the region belongs to high-income and middle-income country group such as Singapore, Malaysia, and Thailand.

M. Alam (✉)
United Nations Environment Programme, Bangkok, Thailand
e-mail: alam31@un.org

P. Sawhney
Asia Pacific Adaptation Network (APAN), Bangkok, Thailand

M. Alam et al. (eds.), *Status of Climate Change Adaptation in Asia and the Pacific*, Springer Climate, https://doi.org/10.1007/978-3-319-99347-8_3

This region is the home of two-third of the world's population of which about 45% lives in urban areas. However, region predominantly is an agrarian society with more than 55% of its total population living in rural areas. Further, large populations' livelihood, more than 80%, is dependent on agriculture. It is estimated that by 2030, urban population will be about 56% and will become 65% by 2050.

The predominant physiographic features in Asia and the Pacific region are (a) river deltas with vast floodplains, (b) Islands and countries with low-lying coastal plains, (c) deserts, and (d) mountain.

The region houses the largest delta along with vast floodplain on earth, i.e. the Bengal delta in South Asia. Other large and important river deltas in the region are Mekong delta, Red River delta, Irrawaddy delta in Southeast Asia, Yangtze River delta, and Zhujiang River delta in Northeast Asia. These deltas are major sources of freshwater supply and freshwater ecosystems, produce large amount of food particularly rice, and directly support livelihoods of a large rural population.

The region is also consisting of Small Island Developing States including 14 Pacific Island Countries, and the largest archipelago, i.e. Malay Archipelago which has over 25,000 islands is situated between the Indian and Pacific Oceans. It includes Brunei, Singapore, East Malaysia, Indonesia, the Philippines, and East Timor. Countries in the region also have large low-lying coastal zones (e.g. Bangladesh, Indonesia, Pakistan, Vietnam). Further, several large cities are in the coastal region (e.g. Ho Chi Minh, Manila, Bombay, Bangkok, Jakarta, Karachi, Shanghai, Tokyo).

The Gobi Desert in Mongolia, Kara Kum (Turkmenistan), and Kyzyl Kum (Kazakhstan) deserts in Central Asia, Taklamakan Desert in China, and Thar Desert in India and Pakistan are major deserts in Asia. Australia also has deserts such as Great Victoria Desert, Great Sandy Desert, and Tanami Desert. The Gobi is the second largest desert in Asia extended from northern China into Mongolia. As the Himalayan Mountains block rain clouds to reach the region, the Gobi Desert receives an average of 7 in of rainfall each year. Among these deserts, the Thar receives up to 20 in of rain per year, primarily during the monsoon period from July to September that help to grow most crops in the area.

The tallest mountain peaks around the globe include the Himalayas; it has several mountainous regions such as the Hindu Kush Himalaya (Bhutan, Nepal, Pakistan, Afghanistan, Bangladesh, Myanmar, part of China), high mountains in Central Asia, and the Tibetan Plateau. The largest mangrove forest, the Sundarbans, is also located in Asia. There are several large water bodies in the region that include Caspian Sea, Aral Sea, and Balkhash Lake.

3.2 Multiplicity of Changes and Vulnerabilities

3.2.1 Climate Change and Stimuli

Change in temperature and variation of rainfall are commonly observed changes across Asia while change in sea level and intrusion of salinity observed in countries with low-lying coastal area. While magnitude of change varies across sub-region and countries within sub-region, vulnerabilities of countries and adaptation needs largely vary by country's readiness to adjust with observed and anticipated changes.

Increased exposure to climate induced water-related extreme events such as floods, flash floods, and droughts and are commonly observed phenomena in Central Asia, Northeast Asia, South Asia, and Southeast Asia specifically countries sharing large river basin such as Ganges-Brahmaputra-Meghna, Mekong, Yangtze River Delta, and Zhujiang River Delta. With warming climate, melting of glaciers and decrease of glacial zone are very vivid in upstream countries in Central Asia such as Tajikistan and Kyrgyzstan, countries in Hindu Kush Himalaya and Tibetan Plateau. Intensive melting of mountain glaciers will initially increase the inflow of water to the rivers, but will drastically decrease the water availability afterwards, as glaciers dry up.

Sea level rise along with inundation and salinity intrusion are also common water-related hazards in South Asia, Southeast Asia, countries in the Pacific as well as some Northeast countries. It is worth noting that in the Pacific, about 70% of the population live in the coastal zone. Countries exposed to Indian Ocean including Bay of Bengal, and Pacific Ocean including South China Sea face frequent typhoons, storm surges, and long-term inundation. In Pacific, the higher category cyclones appear to be becoming more frequent, while the number of cyclones remains stable.

3.2.2 *Impacts and Vulnerabilities*

It is evident that changes in climatic conditions such as increase in temperature, variation of rainfall, hydrological extreme events, and quality and available of natural resources conditions already affecting crop yield and food productivity. Though future impacts will vary across the region, developing countries and least developed countries in many of the sub-regions will experience decline of crop yield (Nelson et al. 2009; IPCC 2014). Shortage of water due to glacier melting will reduce availability of water for irrigation in Central Asia, water-related extreme events such as flood and droughts, salinity in the low-lying coastal areas, and deltaic plains in South Asia, Southeast Asia, and Northeast Asia will adversely affect crop agriculture and food production. It will also significantly alter current capacities of the region's food security and livelihood of small-scale farming communities.

Coastal zone vulnerability would be acute due to the combined effects of climate change, sea level rise, subsidence, and changes of upstream river discharge, cyclone and coastal embankments (South Asia). Four key types of primary physical effects, i.e. saline water intrusion, drainage congestion, extreme events, and changes in coastal morphology have been identified as key vulnerabilities in the coastal area of Bangladesh (South Asia).

Sea level rise poses one of the most widely recognized climate change threats to low-lying coastal areas, islands, and atolls. Over last century, global mean sea level rose at a rate between 1.3 and 1.7 mm/year and over last two decades the rate was between 2.8 and 3.6 mm/year. In the tropical western Pacific where many small island communities exist, rate is up to four times higher than the global average (approximately 12 mm/year).

Future rates of sea level rise are expected to increase coastal flooding, erosion, and saltwater intrusion into surface and ground waters. About 16% coastal land of Small Island States are below 10 m elevation against a global average of 2%. Sea level rise could submerge the lowest parts of the Mekong delta, about 7% of Vietnam's agriculture land and saltwater intrusion could decrease rice yield in Myanmar. Communities have been already displaced due to increasing inundation of low-lying settlement areas in Vanuatu for example. Accelerated coastal erosion and saline intrusion into freshwater sources, decreasing fish distribution and abundance of stock and damage to coral reefs, increased acidification in Pacific.

Cities with extensive port facilities, large-scale petrochemical and energy-related industries are especially vulnerable to risks from increased flooding. With a 0.5 m rise in sea level, the population at risk could be more than triple while asset exposure is expected to increase more than tenfold for cities like Mumbai, Guangzhou, Shanghai, Ho Chi Minh City, Kolkata, Osaka-Kobe, Tokyo, Bangkok, and Dhaka.

There are greater health risks associated with climate change as it will increase the risk of malaria, infectious diseases, and cardio-vascular illness. This will significantly hamper the economic development and pose a risk to human health and security in Northeast Asia, Central Asia, Southeast Asia, and Pacific.

Fragile ecosystems, such as mountains in Hindu Kush Himalayan (HKH) region are suffering from the severe impacts of CC in the forms of degradation of vegetation cover, deterioration of soil, and a reduction in livestock productivity, further impoverishing pastoral communities. The HKH region provides water, ecosystem services, and the basis for livelihoods to a population of around 210.53 million people living in this region. The people in the HKH are highly vulnerable to food insecurity because of their low productivity, subsistence economies, the constraints of terrain, poor infrastructure, limited access to markets, physical isolation, vulnerability to natural risks, and high cost of food production and transportation.

Climate changes occurring in mountain areas are also inducing severe impacts on downstream areas through increasing pressure on natural resources and production systems, and competition for food, accommodation, and income opportunities. In recent years, CC has led to a depletion of the natural resource base across the HKH region (Rasul et al. 2014). This has resulted in a significant loss of ecosystem services, particularly in terms of soil nutrients, water, and biomass, and resultant decline in food productivity (Tiwari 2000). The reduced productivity has affected food security and increased the overall vulnerability of mountain people, as they have become more dependent on food from outside and thus exposed to market fluctuations. Climate change is adding new challenges in terms of water availability and changing temperature conditions (Chap. 10).

3.2.3 *Priority Sectors for Adaptation*

Priority sectors for adaptation to climate change are very closely associated with the highest level of impacts and vulnerability. Sectors prioritized by countries in Asia and the Pacific region are also closely linked to supporting economic development and livelihoods of large populations. Agriculture, water resources, coastal and marine resources have been identified as common priority sectors by all five sub-regions, i.e. Central Asia, Northeast Asia, Pacific, South Asia, and Southeast Asia while fisheries have been identified as a priority sector by South Asia and Pacific countries. Forest and biodiversity have been identified as priority sectors by Northeast Asia, Central Asia, and South Asia while natural ecosystems and human health have been identified as priority sectors by Central Asia, South Asia, and Southeast Asia.

There are sectors prioritized by a country to address their specific needs and vulnerability to climate change. For example, energy and hydroelectric power capabilities in some Central Asia, South Asia, and Southeast Asia will depend on the availability of water resource. Management of water resources for hydropower generation has also been identified as priority.

3.3 Adaptation Goals and Priorities

3.3.1 *Protecting People*

Though expression of adaptation goals varies across sub-regions and countries, the overarching goals are attached to protecting people, development, and ecology from adverse impacts of climate change. Protection of population from water-related extreme events and hazards (South Asia); avoiding impacts on socio-economic development due to water and food shortages, and decreased agricultural production (Central Asia); protect and manage the coasts including human settlement and infrastructure from tropical cyclone, storm surges, and inundation (Pacific and Islands Countries in Southeast Asia, South Asia); and safety of society and protecting land from adverse impacts of climate change (Northeast Asia) are key adaptation objectives.

Further, Pacific Island Countries, several countries in South Asia and Southeast Asia acknowledged evolving future climate risks that may not only impede development but may threaten human security and ecological sustainability. Therefore, efforts to build resilience need to cover not only vulnerable people/communities, but also vulnerable economies/livelihoods, and ecosystems or natural resources.

3.3.2 Ensure Food Security

Not yet set any commonly applicable regional or sub-regional adaptation goal, despite each of the countries in the region has separately expressed its vulnerability concerns to climate change, *food security* has been the common key concern in the national statements on climate change. Food insecurity and inadequate access to food by the poor and the marginal population has been given high attention by South Asia. Throughout South Asia, new food silos are built to maintain a healthy stock of food, only to be utilized during food insecurity. Efforts are now made to build a regional food bank, where food will be mobilized from such silos whenever food shortage is sensed in any of the regional countries.

In the context of food production, it would require protective infrastructure (e.g. to safeguard standing crops), irrigation and water efficiency, reconstruction of irrigation channels and water supply systems, integrated water resource management approach, application of crop diversification and rotation practices, and introduction of new technologies appear commonly suggested adaptation measures.

3.3.3 Enhance Resilience of Infrastructure

Protecting coastal communities from adverse effects of climate change across Asia and the Pacific requires several measures to ensure the safety of populations, security of livelihood assets including ecosystems and their services. Protecting coastal communities also requires to build resilience of key economic sectors and coastal infrastructure such as flood protection, irrigation and drainage, erosion protection and beach nourishment, as well as promotion of alternative livelihood and market development for new products. These measures are not new; rather efficacies of each measures are well tested and understood. In most cases, it would require replication and/or realignment of design with new standard.

3.4 Status of Adjustment and Gaps

Developments in advancing climate change adaptation vary across the countries in the region. Likely reasons include different levels of vulnerabilities, capacities, priorities, resources, and political will. Until the first decade of the present century, none of the South Asian countries has started adaptation activities in a major way, excepting a few trials on Community-Based Adaptation (CBA). However, in the early years of adaptation, several South Asian countries have implemented adaptation projects and have started organizing themselves towards designing and implementing adaptation projects beyond small-scale pilots.

Lack of policies and targeted programmes on climate change and the absence of climate change indicators in the development strategies and poverty reduction documents found evident. Even where exist, proper implementation of climate policies and laws, inadequate institutional and technical capacities, weak coordination between the national institutions and agencies, and insufficient finance and ownership of the government are key barriers for implementation of adaptation measures. Additional constraints include cost-effectiveness of the adaptation initiatives, the level of uncertainty of the risks involved, and the weight of other development priorities.

The current trend reflects the dependency of the legislation of policies and strategies primarily on motivation of the policy makers to pursue the international multilateral process to act. There is a wide disparity in the policy process between the least developed, developing, and the developed countries in Asia. There is an evident need to update and institutionalize climate change adaptation strategies while integrating into existing policies. Overlapping policies in climate change could potentially mislead implementation of strategies. For example, overlapping policies in climate change and agriculture in the Philippines and Korea.

3.4.1 Climate Governance

All countries in Asia and the Pacific region are Parties to the Framework Convention on Climate Change and have focal agency to address climate change domestically and internationally. In the context of formulation and enactment of climate change policy, strategy, action plan, and legislations, it varies by country. All Least Developed Countries (LDCs) in Asia and the Pacific have developed their National Adaptation Programme of Action (NAPA) which has identified country's immediate and urgent needs. Almost all developing countries in Asia and the Pacific region have developed their country-specific climate change strategies and/or action plans. Where a country does not have climate change-specific strategy and/or action plan, for example, Kazakhstan, Kyrgyzstan, and Uzbekistan, exiting development plan, environmental policies, and programmes support countries to cover a range of adaptation activities in key socio-economic sectors.

Though most of the countries in Asia and the Pacific region have strategy and action plan to deal with climate change adaptation, lack of legal provisioning found to be true across the region. Among countries in Northeast Asia, Korea and China are advanced in legislating long-term strategy as well as short-term action plans. In South Asia, Pakistan appears to be the first country to formulate and enact climate change act in Pakistan.

South Asian nations are heavily engaged in climate negotiations and pursue adaptation agenda at international level but legislators in their respective country are reluctant to frame laws and regulations. Part of the problem lies with lack of resources to cater the growing adaptation needs of the poor, which often discourages the political masters not to bind them legally at national level. Involvement of the local government, which is crucial, is lacking to a large extent in Southeast Asia. There is a need of hermonization of adaptation objectives and timeframes set under different policies and plans to meet the adaptation objectives in South Asia.

3.4.2 Institutional Arrangement

Nearly all countries in Asia that lack a national policy and/or act on climate change, coordination among different agencies both vertical and horizontal to design and implement adaptation measures hindered by the weak institutional arrangement and coordination mechanism. This also limits effective engagement with sectoral and sub-national level actors both at policy and strategic level discussion, integration of climate change adaptation into development planning and budgeting system. Absence and/or week institutional mechanism also limits access to resources including financial, technical, and human that may otherwise be efficiently mobilized.

In countries in Central Asia and Southeast Asia, very limited coordinated efforts are going on. There is tendency to implement short-term and fragmented actions, lack of demonstrable on the ground adaptation interventions, and weak or absence of monitoring and evaluation mechanisms. Therefore, it is difficult to assess progress of implementation of climate change policies, strategies, and action plan including their effectiveness. Most of the South Asian countries are also struggling to come up with strong institutional arrangement towards implementing climate change adaptation measures in an effective manner. For example, Bangladesh has established its own climate change trust fund through budgetary allocation since 2009–2010 fiscal year, but little institutional mechanism has been put in place to ensure a coordinated adaptation action, monitoring results, and feedback and learning from implementation as well as to ensure value for spending.

While a well-placed institutional framework of the government to bridge the communication and coordination gap between sectoral ministries and government agencies exists, there are countries in the region that have taken steps and have established strong coordination mechanism. Three out of four countries in Northeast Asia have established cross-ministerial committee, while Korea designated an institution to support the ministry for climate change adaptation. Republic of Kyrgyzstan has established Climate Change Coordination Commission which is chaired by the First Prime Minister. Similar coordination mechanism has also established by the Government of Philippines. In the Pacific, most of the climate change focal points are positioned within their central government agencies while operational focal points tend to be housed in line ministries.

3.4.3 Integration and Mainstreaming

Among five sub-regions, the Pacific is the first region in the world that decided on a single integrated regional strategy to address climate change and disaster risk management. Most Pacific Island Countries integrate climate change adaptation and disaster risk management into their national policies, development strategies, and legal and governance frameworks. Countries in South Asia are making greater efforts towards integration of climate change adaptation with disaster risk reduction and development. However, very limited progress has been made to change design standards particularly in designing infrastructures. In many cases, climate change-related hard data is not collected or analysed, which might be a prerequisite to change design criteria. Therefore, climate resilience design of infrastructure often left to the intuition of engineers and value has been considered arbitrarily.

There are noticeable differences in the approach of Northeast Asian countries towards mainstreaming adaptation. It ranges from consideration of adaptation in 5-year national development plan, i.e. China while Korea and Mongolia are using both national development plan, sectoral implementation plan. Though Japan has organized its National Adaptation Action Plan around a top-down concept of implementation, it explores bottom-up approaches to adaptation to CC. An integrated concept of "hard" and "soft" measures has also proposed in Japan to manage adaptation.

Except Kyrgyzstan, rest Central Asian countries lack mechanism to integrate climate change adaptation into existing policies and regulations of sectoral ministries and agencies. Though Kazakhstan does not have a targeted framework document on climate change adaptation, it acknowledges mainstreaming of climate adaptation into the sector development planning. The Government of Kyrgyzstan advanced integration of climate change into National Strategy of Sustainable Development for 2013–2017 and sector-based integration (healthcare, natural disasters and emergencies, forest and biodiversity, water, and agriculture). One of the strongest sides of the Kazakh government in climate change adaptation is replication of technologies and to provide enabling policy tools for the farmers and entrepreneurs.

Integration of climate change adaptation at local level is constrained by bureaucracy and political barriers. Unclear delineation of functions of different ministries and agencies in many policies hinder integration as well. Though adaptation at the local level has been recognized as pivotal, major policy documents do not provide a clear and coherent institutional framework for implementing climate change adaptation activities at the local level. Further resources and capacities are still centralized and the means to avail the services often fail to reach the most vulnerable. There is also issue of weak accountability and transparency as well as the lack of or inadequate dialogues between the authorities and recipient of adaptation measures such as farmers.

3.4.4 Finance

Almost all countries in Asia and the Pacific region have accessed climate finance for implementing adaptation interventions at different levels from different sources. These include funds established under the framework convention on climate change, development banks, bilateral development partners, philanthropic organizations, etc. Establishment of new national level climate funds (e.g. Bangladesh), widening scope of existing environment-related funds (e.g. Bhutan) as well as earmarked financial support for adaptation programme (e.g. China and Korea), budgetary allocation (e.g. Japan) appears an emerging trend in the region. Further financial assistance was also given in the fields of agriculture, water resources, coastal ecosystem conservation, and public health.

Financing adaptation measures from private sector is often limited and mostly concentrated around risk insurance. A common issue in every country in the region is how to make private sector engaged in adaptation. It has been recognized that the increase of awareness and interest on subject is difficult when a business or sector is not affected directly by climate change.

While in-country capacity to access finance, adequacy of finance, and predictability of long-term financing are common issues, insufficient and underutilized mechanisms to generate alternative sources of climate financing are also common. Social and economic development priorities also hinder allocation of national level funds for adaptation projects as well as building long-tern institutional capacity to support adaptation planning.

3.4.5 Implementation Capacity

Possessing capacity at the systemic, organizational, and individual levels is very much essential for a smooth implementation of climate change policy and measures on adaptation. Considering different climate change adaptation projects in sectors and types of adaptation technologies being implemented by countries in the region, one can argue that the countries have experiences of implementation of adaptation projects and thus have capacity. However, implementation capacity and knowledge are limited and mostly possessed by couple of ministries and sectors while climate change cuts across all sectors and society (Central Asia, Mongolia in Northeast Asia, South Asia).

Implementation capacity gaps in the region include lack of institutional and administrative capacity especially at the local/community level. Local government authorities are often unable to translate national policies into local level plans. There is lack of technical skills/expertise in climate science, modelling, and socio-economic impact assessment. Low level of understanding about uncertainties and limitations of projections/scenarios also work as a barrier to design and implement adaptation measures.

Implementation capacity is not homogenous even in countries within a sub-region. For example, quality of implementation in India and Sri Lanka appears to be far better than in Bangladesh and Nepal. India has a proven edge in South Asia regarding absorption capacity as human capital is well equipped to tackle the tasks ahead of them. Yet, India wishes to enhance its institutional and human capacity further to address the issues concerning climate change.

While large number of countries have deficiency in analytical skills and human capacity within sectoral ministry and agencies, lack of monitoring and evaluation system for reporting results or achievement (Central Asia, Southeast Asia), couple of countries in Southeast Asia have set up specific indicators to measure effectiveness of the implementation of planned adaptation actions (Cambodia, Philippines).

3.4.6 Technologies

Technologies for adaptation usually refers to an equipment, technique, practical knowledge, or skills that have the ability to reduce vulnerability to climate change. Discussion on technologies for adaptation has been broadened here to cover technologies required during different stages of adaptation planning such as to develop information and raise awareness, plan and design adaptation strategies, implement adaptation strategies, and to monitor and evaluate performance.

Countries in the region know their technological needs and many of them have already assessed the needs under Technology Needs Assessment exercise. Many countries are implementing technologies for adaptation to reduce vulnerability of key impacted sectors and people. Types of technologies currently being applied in different vulnerable sectors have commonalities such as water saving techniques for agriculture that include micro and drip irrigation techniques, alternate furrow, double-sided furrow application, surge flow, laser levelling, etc. (Central Asia). Countries are promoting climate resilience agriculture technologies such as switching to less water intensive crops, drought-resistant crops, crop rotation, zero tillage (Central Asia, South Asia, and Southeast Asia), and saline tolerant crop varieties (South Asia and Southeast Asia).

Several countries in the region have recognized the importance and have made significant investment in early warning technologies for a functioning platform for public meteorological services, improved its climate observation network, and heightened the timeliness of meteorological services (e.g. China), reliable weather forecasting systems to predict the intensity and speed of tropical cyclone (e.g. Philippines, Bangladesh, India), early warning systems for floods, flash floods drought advisories (Bangladesh). Similarly, Japan has developed one of the most advanced climate monitoring systems in the world.

Application of technologies for better understating of vulnerability at local level and support adaptation planning is also emerging in the region. For example, Korea has developed vulnerability maps at a small scale (provincial and municipal) and built an online service platform called Vulnerability Assessment Tools to Build

Climate Change Adaptation Plan (VESTAP). Similarly, Southeast Asian countries are using the Geographic Information System (GIS) to map out vulnerabilities of flood, drought, and sea level rise. In Singapore, research and technology development are being harnessed to improve and augment water supply (i.e. NEWater, desalination).

While countries in Asia and the Pacific region are implementing different technologies to support adaptation planning and interventions, challenges do exist in several areas. For example, technologies like micro-irrigation techniques and laser levelling require large investments and are too expensive to use widely. Difficulties also occur in downscaling climate scenarios due to limited or lack of availability of climate data, difficulty in accessing data, and high uncertainties and limitations of climate scenarios/simulation.

Countries in the region require innovation and scale up of technology application to trigger transformational adaptation. Adaptation research is yet needed to understand what society needs to adapt to in concrete terms, to identify available options, and to understand how adaptation can be effectively promoted and implemented. There are countries in the region advancing in research and innovation. Bangladesh, India, and Pakistan are ahead of other South Asian countries in adaptation technology innovation, development, and adoption. There are countries in the region who have established institutions to advance science and technologies. For example, the Department of Science and Technology (the Philippines), the Thailand Institute of Scientific and Technological Research (Thailand), and the State Ministry of Research and Technology (Indonesia).

3.4.7 Decision-Making Tool and Institutionalization

Both technical and financial capacity to apply robust decision-making tools and internalization of capacity within the government institutions varied between Least Developed Countries and Developing Countries. Countries with higher economic strength usually possess higher technical capacity. The analytical rigour and output qualities largely vary among countries in South Asia. The smaller countries mostly depend on external support, both technical and financial, to complete vulnerability assessments.

Limited or incomplete climate data, lack of downscaled climate projections, uncertainties in climate simulation, lack of locally relevant methodologies and tools are also common issues in most of the countries. All these are valuable inputs for well-informed adaptation planning. Absence of validated climate modelling results often leads to use of gross approximations and assumptions that compromise robustness of decision. Further, countries have not developed yet any noticeable institutional mechanism and process to undertake for periodically carrying out analysis and to provide updated information on vulnerability and adaptation needs.

3.4.8 Awareness and Access to Information

Status of awareness and access to information vary by sub-region and countries across region. Most climate change adaptation projects often include awareness raising activities such as targeted workshops, seminars, and public awareness events. Target audience of awareness raising ranges from policy makers to local communities. There is no systematic assessment of awareness; it appears that general awareness on climate change has increased over time. However, perception still prevails that climate change is an environmental challenge rather a development issue. Therefore, it should be tackled by respective "environmental" institutions.

Information related to vulnerability and sectoral impacts to climate change are available in most cases but dissemination to local people is limited. Further, often dissemination does not use the forms and languages which are palatable to the local mass (South Asia). Lack of scientific research and an insufficient awareness about the need for adaptation lead naturally to a limited belief for adaptation need (China). Lack of awareness and understanding of climate risks and their impact on development (Central Asia) as well as limited access to information (S Asia), access to real-time monitoring data as well as long-term databases (NE Asia) remain as significant barriers to design and implement adaptation measures. One of the biggest challenges in Central Asia and South Asia is lack of media engagement in awareness raising.

In terms of long-term impact of awareness raising activities, development of educational materials and teaching at elementary schools is an excellent example of Korea's efforts to improve general public's understanding of adaptation.

Climate change for developing countries in the region can be defined as additional development challenges as it cuts across all sectors of its development. Almost all developing countries in the region are struggling to meet their development needs such as poverty alleviation, education, health, managing natural resources to support livelihoods, and economic sectors. It can also be seen as opportunities to make its future development climate resilient and low carbon economic development. Many cities which are the engine of economic growth are also vulnerable to climate change and therefore require higher attention to address impacts of climate change like agriculture and water resources.

References

IPCC (2014) Asia. In: Barros VR, Field CB, Dokken DJ, Mastrandrea MD, Mach KJ, Bilir TE, Chatterjee M, Ebi KL, Estrada YO, Genova RC, Girma B, Kissel ES, Levy AN, MacCracken S, Mastrandrea PR, White LL (eds) Climate change 2014: impacts, adaptation, and vulnerability. Part B: regional aspects. Contribution of working group II (authors: Hijioka Y, Lin E, Pereira JJ, Corlett RT, Cui X, Insarov GE, Lasco RD, Lindgren E, Surjan A) to the fifth assessment report of the Intergovernmental Panel on Climate Change. Cambridge University Press, Cambridge, pp 1327–1370

Nelson G, Rosegrant M, Koo J, Robertson R, Sulser T, Zhu T (2009) Climate change impacts on agriculture and costs of adaptation. International Food Policy Research Institute, Washington DC

Rasul G, Hussain A, Khan MA, Ahmad F, Jasra AW (2014) Towards a framework for achieving food security in the mountains of Pakistan, ICIMOD working paper 2014/5. ICIMOD, Kathmandu

Tiwari P (2000) Land-use changes in Himalaya and their impact on the plains ecosystem: need for sustainable land use. Land Use Policy 17(2):101–111

Chapter 4
Status of Climate Change Adaptation in Central Asian Region

Nailya Mustaeva and Saniya Kartayeva

Abstract Central Asia is highly vulnerable to adverse impacts of climate change and extreme weather events. Countries have identified water, agriculture, energy, human health, natural ecosystems, biodiversity, and forests as highest priority sectors for adaptation. This chapter examines current efforts and status of climate adaptation and identifies critical gaps for enhancing adaptation actions in Central Asian countries. It analyses the sub-region's capacity to cope with climate impacts, making the links to the existing legislative basis and national policies, institutional arrangements, access to finance and decision-making process. It reveals that they share similar challenges to address climate change adaptation needs and development priorities.

Keywords Climate change · Adaptation · Resilience · Mainstreaming · Integration · Central Asia

4.1 Introduction

Central Asia is in the heart of Eurasia with continental arid and semi-arid climate with hot and cloudless dry summers, and moist and warm winters in the south and cold winters with severe frosts in the north (Lioubimtseva and Henebry 2009). It stretches from the Caspian Sea in the west to China in the east and from Russia in the north to Afghanistan in the south, presenting a big variety of landscapes—deserts, and steppes and high mountains. The region is represented by five countries, including Kazakhstan, Kyrgyzstan, Turkmenistan, Tajikistan, and Uzbekistan (Fig. 4.1).[1] Total population of Central Asia is more than 67 million (2014) inhabited in the area equalling to 3.98 million sq. km. Central Asia to the most extent is a landlocked region with no direct access to the ocean (WB 2015). The main water bodies in the region include Caspian Sea, Aral Sea, Balkhash Lake, small and

[1] http://www.zoinet.org/web/sites/default/files/large-maps/Central_asia_shading.png

N. Mustaeva (✉) · S. Kartayeva
The Regional Environmental Centre for Central Asia (CAREC), Almaty, Kazakhstan

M. Alam et al. (eds.), *Status of Climate Change Adaptation in Asia and the Pacific*, Springer Climate, https://doi.org/10.1007/978-3-319-99347-8_4

Fig. 4.1 Map of Central Asia

medium size rivers and lakes. Since the region is in the centre of Eurasian continent and without open access to water, availability of water resources and its effective management are among top priorities.

Central Asia is highly vulnerable to climate change. The countries of the region have the highest index of vulnerability and lowest index of coping capacities in the whole Eurasian continent (Barbone et al. 2010). Increase in near-surface temperature, deviations in atmospheric precipitation, melting of mountain glaciers and its effects on water regime and runoff are the main indicators of climate impacts. Various expert assessments indicate that Central Asia lost almost 35.5% of its glacier in volume during 1901–2000, and 11% between 1980 and 2011 (Giesen and Oerlemans 2013; Hijioka et al. 2014; Marzeion et al. 2012). With climate warming, glaciers melting, and degradation will continue with significant impacts on water runoff and development sectors of economy, human health, and natural ecosystems. The region is expected to become warmer in the coming decades and increasingly arid in the western parts of Turkmenistan, Uzbekistan, and Kazakhstan (Lioubimtseva and Henebry 2009; IPCC 2014). Scenarios of temperature and precipitation change in Central Asia by 2030, 2050, and 2085 are shown in Table 4.1.

Undoubtedly, projected changes in precipitation and temperature, as well as melting of glaciers and its link to water availability for irrigation and hydropower generation, will most likely affect socio-economic development of the region. It will become more vulnerable to climate-induced disasters, such as flash floods and

Table 4.1 Scenarios of temperature (T) and precipitation (P) change in Central Asia

Country	2030	2050	2085
Kazakhstan	T: +1.4 °C P: +2.0%	T: +2.7 °C P: +4.0%	T: +4.6 °C P: +5.0%
Kyrgyz Republic	–/–	–/–	T: 6.1 °C; 4.6 °C P: −2.2%; +8.3%
Tajikistan	T: +0.2–0.4 °C P: +2.0%	–/–	–/–
Turkmenistan	–/–	–/–	T: +4.6–5.5 °C P: −17.0–56.0%
Uzbekistan	T: 1.2 °C P: 2.0%	T: 2.2 °C P: 4.0%	T: 3.3 °C P: 3.5%

Source: CAREC (2011)

droughts, water, and food shortages, and decreased agricultural production (wheat, rice, and cotton). This will ultimately impact people living in Central Asia and make them more vulnerable to climate risks. National communications state that key health risks associated with climate change such as increased risk of malaria, infectious diseases, and cardio-vascular illness will primarily affect old people and children.

Melting of glaciers and decrease of glacial zone in the region are other vivid examples of climate change. Upstream countries, like Tajikistan and Kyrgyzstan, being the main glacial knots and water towers in the region, nowadays are experiencing intensive melting of mountain glaciers which will initially increase the inflow of water to the rivers, but will drastically decrease the water availability afterwards, as glaciers dry up. For example, the total glacial area of Tajikistan reduced from 6 to 5% of the total country's territory during the last decades (GoT 2014), and resulted in the decreased river runoff for the Aral Sea basin. Estimates show that at present it equals to 53 km^3, which is 4 km^3 less than it was 50 years before (GoT 2014).

Climate scenarios suggest that this tendency will continue in the forthcoming century. IPCC findings show that within the next 20 years the flow of main rivers such as Amudarya and Syrdarya will be likely reduced by 20–30% (IPCC 2007). Kyrgyzstan's Second National Communication (SNC) estimates that by 2025 decrease of surface water flow will be around 43.6–88.4% from the volume of flow in 2000 (GoKR 2009). Consequently, downstream countries will face a shortage of water needed to maintain economic development. In Uzbekistan, total deficit of water in 2005 was 2 km^3 and will be amounted to 7 km^3 by 2030 and 11–13 km^3 by 2050 (GoUz 2008). Turkmenistan's industry and agriculture heavily rely on water from Amudarya. Almost 90% of its runoff is used for agriculture. It is estimated that demand of water for irrigation, which is about 25 km^3, may not be met in future in case no adaptation measures are taken today (GoTr 2010).

4.2 Key Impacted Sectors and Adaptation Priorities

Since the countries of Central Asia share the similar climate and environmental challenges as well as development priorities, the defined priorities towards adaptation also have many commonalities. For example, water resources, agriculture, energy (hydropower generation), human health, natural ecosystems, forests, and biodiversity are considered by all five Central Asian countries in their national communications as the most vulnerable sectors to climate risks, and should be treated first and foremost. At the same time, extreme weather events and climate-induced disasters are considered as cross-cutting for every sector (Table 4.2). Undoubtedly, the level of impacts might vary from one country to another, but the overall scope of the needed adaptation measures is mostly based on global and regional forecasts and climate scenarios.

The water sector is the most vulnerable to climate change and is strongly dependent on the magnitude of glaciers melting. Estimates show that glaciers will be significantly reduced by the end of this century, leading to temporary increase of water flow of the main rivers of the region. In the long term, future water flow will decrease and significantly impact the irrigated areas (GoT 2008). Turkmenistan, Kazakhstan, and Uzbekistan, which are located downstream of the main rivers will be affected the most. SNC Uzbekistan (GoUz 2008) estimates that the water availability of Amudarya River may decrease by 10–20% by the end of twenty-first century. The annual water discharge of Syrdarya River in Kazakhstan will most likely drop from 7756 million cubic metres to 6528 million cubic meters, about 16% reduction, by 2020 (IWRM 2011). Hence, key adaptation objectives in this sector, which are commonly suggested by the countries in their national communications, envisage (1) improve access to clean drinking water and increase effectiveness and reliability of irrigation water systems; (2) promote Integrated Water Resource Management (IWRM) principles; (3) integrate advanced methods of water saving consumption in different sectors of economy, especially in agriculture sector (drip, micro-spray, etc.); and (4) use of alternative sources of water for irrigation such as low-mineralized collector-drainage and groundwater.

Agriculture in Central Asia plays an important role in the economy and contributes between 10% and 40% of gross domestic product (GDP), with more than 40% of population involved in the sector (CAREC 2012). Climate change will dramatically impact agricultural development and food production. Less glacial runoff and higher temperatures in summer will put pressure on water resources, and prolonged periods of heat stress could reduce the productivity of many types of crops and livestock. For example, in Kazakhstan, projections show that by 2050 production of cereals will drop by 37–48% as compared to 1971–2010 (GoK 2013). In Turkmenistan, water demand for crops like cotton, wheat, and vegetable will increase from 7347 to 8271 m^3/ha, if temperature rises between 0.1 and 1.9 °C (GoTr 2010). With fewer resources for agriculture, rising food prices could put many at risk of hunger and undernourishment (WB 2014). Shifting common crop production practices by introducing new drought and climate-resistant varieties of

Table 4.2 Vulnerability of key development sectors to climate change in Central Asia

Priority sectors	Kazakhstan	Kyrgyzstan	Uzbekistan	Tajikistan	Turkmenistan	Observed changes
Water sector	√	√	√	√	√	Surface area of the Aral Sea Basin fell by about 57%, the volume by about 80%, and depth by 64% in the past four decades (WB 2014)
Agriculture	√	√	√	√	√	Severe droughts in 2000–2001 leading to 112,600 ha of cereal loss in Tajikistan and US$ 50 million loss in Uzbekistan (Thurman 2011)
Energy (hydropower generation)		√		√		In the past 45 years, the siltation of the Nurek reservoir might have reached >100 m or 45% of the total dam's dead volume (Sirojiddinov 2012)
Human health	√	√	√	√	√	Average number of deaths due to climate-induced disasters in Kyrgyzstan increased from 61 in 2003–2007 to 281 in 2008 (MoH 2011)

(continued)

Table 4.2 (continued)

Priority sectors	Kazakhstan	Kyrgyzstan	Uzbekistan	Tajikistan	Turkmenistan	Observed changes
Natural ecosystems, biodiversity, and forests	√	√	√	√	√	Desertification now affecting 66% of Kazakhstan territory (WB 2013)
Climate-induced natural disasters and extreme weather events	√	√	√	√	√	During 1997–2001, due to extreme weather events, about 3.6 thousand km of roads and 500 bridges have been destroyed and damaged in Tajikistan (GoT 2002)

Source: GoK (2013), GoKR (2009), GoT (2014), GoTr (2010), GoUz (2008)

crops, advanced water saving techniques for irrigation, maintenance of land-reclamation activities, sustainable practices of land and pasture management, and improvement of agrometeorological forecasts and early warning systems are common priorities of the countries of the region (GoK 2013, GoT 2008, GoTr 2010, GoKR 2009, GoUz 2008).

The future of the region's energy and hydroelectric power capabilities will also depend on the availability of water. According to IPCC *"the projected decrease in precipitation in Central Asia will be accompanied by an increase in the frequency of very dry spring, summer and autumn seasons. Changes in seasonality and amount of water flows from river systems are likely to occur due to climate change. Changes in runoff of river basins could have a significant effect on the power output of hydropower generating countries like Tajikistan, which is the third highest producer in the world"* (IPCC 2014). With only 8% of the total regional potential, hydropower could play a major role in development of Central Asian countries. Melting glaciers generally lead to higher water availability to generate power, but with changes in runoff patterns less water might be available for power generation in summer (WB 2014). In addition to changes in runoff, hydropower sector will also face a risk of higher siltation due to climate-induced erosion and disasters, particularly floods and landslides. For example, annual average rate of siltation in Tajikistan's Nurek dam is from 2.5 m up to 6 m. Over the past 3 years, siltation of the reservoir increased by 7 m. According to the expert assessments, the rate of the siltation is turned to be higher than initially expected (Sirojiddinov 2012; Verhoturov 2015).

Human health is also recognized as a key sector that will face climate-induced threats in Central Asia. Direct impacts of climate change on human health are associated with heat stress, death or injures during extreme weather events and climate-induced disasters like floods and mudflows. It is projected that May and September will become the most uncomfortable period of the year when extremely high temperatures will affect people's health and may increase a risk of cardiovascular disease and malaria (GoUz 2008). It is expected that by 2030, extreme thermal loads or stresses on humans may increase by 6–18% compared to a basic norm of 34–100 Wt (GoUz 2008). Indirect impacts of climate change to human health are observed through an increased rate of infectious diseases, mainly caused by poor quality of drinking water. For instance, when floods hit Tajikistan in 2010, the capacity of the Kulyab water treatment system dropped to 30%. Affected people were temporarily re-settled in tents in the city's stadium, which did not meet favourable living conditions with basic sanitary services and hygiene. The risk of infectious diseases increased (Mustaeva 2013). The countries of Central Asia have a common vision towards adaptation to human health, which include the development of monitoring system to control the risk of infectious diseases due to climate change, improvement of early warning system for extreme weather events and their impact on human health and security, improvement of drinking water quality and access to water, raising awareness about impacts of climate change, and appropriate adaptive measures.

Natural ecosystems, biodiversity, and forests are in particular vulnerable to climate change. The countries of Central Asia will observe shifts of ecosystems to the north, extinction of endangered species, reduction in range and degradation of wild ecosystems outside protected territories, reduction of areas with juniper and tugai forests, and decrease in forest productivity. Unfortunately, the role of forests is underestimated in the region, despite that it could lead to a win-win solution and increase adaptation and mitigation efforts of the countries. Reforestation and afforestation measures could help not only to stabilize soil and keep water moisture, increasing land productivity, improving local livelihoods, protecting fragile natural ecosystems and biodiversity, reducing the growing risk of floods and mudflows while contributing to the regulation of CO_2 in the atmosphere. Other adaptation measures, which have been suggested in the Central Asian National Communications, include the need for the development and maintenance of the environmental monitoring system for rare and endangered species and conduction of additional research and studies on flora and fauna and its implications of climate change in the region.

The countries of Central Asia are highly prone to the risk of extreme weather events (EWE), which are mostly associated with hydro-meteorological phenomena (drought, heavy rains, thunderstorms, hail, and dust storms), significantly hamper the economic development, and pose a risk to human health and security. The total cost of loss and damages that are caused by floods, droughts, and landslides in 1993–2011 in Central Asia has been estimated up to more than US$ 2.2 billion, with Tajikistan being accounted to US$ 1.7 billion (Akhmetkaliyeva 2015). In Kazakhstan, between 2002 and 2011 the loss from extreme weather events was about UD$ 68.6 million (GoK 2013). In Kyrgyz Republic, 95% of the settlements are situated along rivers and water bodies, which make these territories vulnerable to floods and mudflows (GoKR 2009). The forecasts for Uzbekistan indicate that the country will face up to 19–24% increase in floods and mudflows by 2030–2050 compared to nowadays. The number of days with extremely high air temperatures (>39 °C) will lead to the 9–15% reduction of crop yields (GoUz 2008). In the region, Tajikistan remains one of the countries, dramatically exposed to climate-induced disasters, which have a sound effect on economic development. Tajikistan faces up to 500 of different natural disasters every year, and annual losses and damage fluctuates from US$ 20 to US$ 100 million (Mustaeva et al. 2015).

It is evident that in Central Asia priority sectors to adapt to climate change are those associated with the highest level of vulnerability to climate impacts and risks. They cover water resources and sustainable water use and management, agriculture and food security, energy, and hydropower generation (particularly for Tajikistan and Kyrgyzstan). Human health and natural ecosystems are also recognized as key sectors, where adaptation measures are urgently needed. Disaster risk management and extreme weather events are considered as cross-cutting concerns around all socio-economic sectors. Key interventions to support climate change adaptation include (1) revisiting the development and sector-based policy and legal frameworks to include climate change adaptation measures; (2) accommodate economic instruments (e.g. incentives, insurance programmes); (3) increase access to finance to effectively mainstream climate change adaptation measures to the sector-based policy frameworks and consider domestic allocations; (4) strengthen the institutional and technical capacity of line ministries and governmental institutions, and (5) increase information exchange, awareness raising and networking on climate change, using mass media for outreach to the general public to a regular basis.

The Table 4.3 below summarizes key priority sectors and suggested adaptation measures, which are common for all countries in Central Asia. The summary is based on information and figures available in Central Asian National Communications.

4.3 Current Trends of Adaptation and Resilience

All five countries of Central Asia are Parties of the UNFCCC and have established relevant legal and regulatory frameworks to meet their commitments under the convention. On a regular basis, the governments develop their National Communications (NCs) and Biennial Update Reports (BURs) on climate change, where they report on greenhouse gas emissions and sinks, vulnerability and adaptation, climate mitigation actions and awareness rising. In 2015, Kazakhstan, Kyrgyzstan, and Tajikistan prepared their Intended Nationally Determined Contributions (INDC) and stated their climate pledges covering both adaptation and mitigation.

Over the past few years, the region of Central Asia has enjoyed an extended support from the international community to cope with climate change. On November 3, 2015, the World Bank Board of Executive Directors approved an allocation of US$ 38 million from the International Development Association (IDA) to finance the first phase of the Climate Adaptation and Mitigation Program for Aral Sea Basin (CAMP4ASB). This approved financing includes US$ 9 million for Tajikistan, US$

Table 4.3 Priorities for adaptation measures in Central Asia

Priority sectors	Climate change impacts	Adaptation measures as stated in National Communications	Proposed common adaptation measures for all sectors
Water	• Melting of glaciers ⇒ increased water runoff ⇒ increased risk of natural disasters (flash floods) in a short-term future • Decrease in water runoff ⇒ increased risk of water deficit for all development sectors in a long-term future • Decrease in water runoff ⇒ increased risk of natural disasters (droughts, forest fires)	• Improve access to clean drinking water and increase the effectiveness and reliability of irrigation systems • Promote integrated water resource management (IWRM) principles • Integrate advanced methods of water saving consumption in different sectors of economy, especially in agriculture sector (drip, micro-spray, etc.) • Use alternative sources of water for irrigation such as low-mineralized collector-drainage and groundwater	1. Revisit the development and sector-based policy and legal frameworks (programmes, action plans, mandates, and regulations) to include climate change adaptation measures 2. Consider economic instruments and stimuli in the sector-based policy frameworks considering climate change (e.g. incentives, insurance programmes) 3. Increase access to finance resources to effectively mainstream climate change adaptation measures to the sector-based policy frameworks, consider the state budget resources 4. Build the institutional and technical capacity of line ministries and governmental institutions on climate change impacts 5. Increase the information exchange, awareness raising and networking on climate change
Agriculture	• Decrease in water runoff ⇒ decline of agricultural yields; land degradation, increased aridity; a high risk to food security and nutrition • Decrease of forage and increased risk of natural disasters ⇒ risk to a livestock	• Use advanced technologies of land management (no-till farming, plant-growing diversification, drought-tolerant crops and cattle, grassland rotation, rational use of pastures, phyto-land reclamation work, organic land management) • Improve the quality of agrometeorological forecasts for agricultural needs (early warning system, extreme weather events, etc.) • Introduction of diversification programmes and cultivation of drought-resistant agricultural crops	
Energy (hydropower generation)	• Change of river flow regime and water runoff ⇒ power generation; • Increased risk of natural disasters ⇒ siltation of water reservoirs; damage of HPP infrastructure ⇒ decrease of energy production	• Improve access to energy sources especially in rural areas (e.g. off-grid techniques and renewable sources) • Enhance hydropower capacity: construct new energy generation units and maintain existing infrastructure • Diversification of energy sources (other types of energy: solar, wind, biogas) • Increase energy efficiency on a demand side	
Human health	• Heat waves ⇒ increased mortality rate among the most vulnerable group of people • Climate-induced disasters ⇒ increased risk of infectious diseases	• Enhance early warning and monitoring system of climate-induced factors and human health • Enrich scientific and evidence base of climate change and its impacts on human health (vulnerable and marginalized population) • Conduct preventive and health protection measures in light of climate-induced disasters	
Natural ecosystems, biodiversity, and forests	• Climate warming ⇒ change in flora and fauna, extinction of rare species • Climate warming ⇒ increased rate of pests and forest fires	• Maintain the specially assigned natural corridors between protected areas • Develop and maintain the environmental monitoring system (rare and endangered species) • Increase the country's carbon sequestration capacity through reforestation and afforestation measures	
Climate-induced natural disasters and extreme weather events	• Change in water runoff ⇒ increased risk of climate-induced disasters ⇒ impact on all sectors of economic development, natural ecosystems, and human health	• Improve the disaster risk management system in terms of prevention, preparedness, and response to climate change impacts • Enhance hydro-meteorological monitoring and early warning systems • Set up and maintain a regular insurance system and social safety nets	

14 million for Uzbekistan, and US$ 15 million for regional activities. CAMP4ASB project aims to enhance regionally coordinated access to improved climate change knowledge services for key stakeholders (e.g. policy makers, communities, and civil society) in participating Central Asian countries as well as to increase investments and capacity building that, combined, will address climate challenges common to these countries (WB 2010, 2015).

At present, out of all five countries of Central Asia, only Kyrgyzstan, Tajikistan, and Turkmenistan have set target to reduce climate-induced vulnerability through implementation of adaptation measures by 2030, using internal and external resources (GoT 2015; GoKR 2015; GoTr 2015). Kyrgyzstan's INDC estimates that in case of non-action, economic loss arising from climate change (energy, agriculture, water resources, emergencies, health, forests, and biodiversity) will be about US$ 1230.8 million by 2030. Similarly, Turkmenistan's INDC also estimates up to 20% decline of crop production by 2030 which will result substantial economic loss worth of US$ 20.5 billion. Tajikistan's INDC mostly speaks about priorities and types of adaptation measures. By its own domestic resources[2], the Government of Tajikistan plans to reduce overall adverse impacts of extreme weather events and climate change by ensuring the implementation of the relevant sector-based programmes and plans of actions with relevance to climate change.

Both INDCs and NCs of all countries in Central Asia commonly identified water, agriculture, energy, human health, natural ecosystems, biodiversity, and forests as the highest priorities for adaptation. At the same time, the overall objective of adaptation is to maintain climate-resilient socio-economic development considering growing needs of the population to the most extent. With climate change, risks for a resilient future will require more proactive actions and coordination of existing and planned initiatives on sustainable development. The section provides a summary of existing trends in policy, institutions, technology, finance, and mainstreaming principles in every country of Central Asia, and highlights best practices and strong sides of adaptive capacity, which can be replicated and used for enhanced measures.

4.3.1 Kazakhstan

Kazakhstan is a candidate for Annex-I membership in the UNFCCC and since 2009 holds a special status as an Annex-I Party to the Kyoto Protocol. The Kazakh GHG inventory leads Central Asia in terms of quality and completeness, and the country's emissions modelling uses advanced tools. In 2014, the country published its first biannual update report as part of its special status reporting obligations and in 2015 submitted its INDC with the pledge to reduce 15% of greenhouse gas emissions by 2030 from 1990 GHG level (GoK 2015). Kazakhstan is a pioneer in emissions

[2] Domestic resources refer to funds within the budget allocated to ministries and institutions for the relevant period.

Table 4.4 Technology needs assessment in Kazakhstan's water sector

Sector	Technology	Scale	Perspective
Water	• Improvement of legislation in drinking (municipal) water supply and wastewater disposal sub-sector	Large scale	Short term
	• Extreme events prevention technology		Long term
	• Renovation of networks and facilities of water supply and wastewater disposal organizations		Medium term
	• Leakage control, detection, and elimination of leakages in pipe networks		Short term
	• Catchment of rainwater and melt water, construction of tanks and reservoirs		Long term
	• Metering of water for irrigation and watering	Large scale	Medium term
	• Reconstruction and renovation of hydro-technical constructions—HTC (water reservoirs, dams etc.), irrigation systems, and networks		Long term
	• Drip irrigation technology	Medium scale	Medium term

Source GoK (2010a, b)

trading in Central Asia (Pawlowski 2012; WB 2013; Zoi Network 2016). In 2013, it launched the national emissions trading scheme (ETS) with an aim to combat climate change and reduce industrial greenhouse gas emissions cost-effectively.

Although the Government of Kazakhstan does not have a targeted framework document on climate change adaptation, it acknowledges mainstreaming of climate adaptation into the sector development planning. In 2013, Kazakhstan adopted the concept for transitioning to a green economy, with the aim to have 50% of alternative and renewable energy share in the total energy balance in a long-term perspective (WB 2014). Concept's action plans foresee measures on sustainable management of water resources, productive land management and food security, clean energy with the use of new "green" technologies. One of the strongest sides of the Kazakh government in climate change adaptation is to extend the replication of technologies and provide enabling policy tools for the farmers and entrepreneurs. For example, the country is progressing well in incentivizing agricultural sector to use water saving techniques and cultivating drought-resistant crops (GoK 2010a, b). Table 4.4 summarizes an overview of needed technologies on adaptation in the water sector of Kazakhstan.

4.3.2 *Kyrgyzstan*

The Kyrgyz Republic recognizes the importance to combat climate change and undertakes serious efforts for an effective policy making. Key strategic documents for a climate-resilient development are "National Sustainable Development Strategy of the Kyrgyz Republic for 2013-2017" and the "Program of the Kyrgyz Republic on Transition to Sustainable Development for 2013-2017". In 2013, the Government established Climate Change Coordination Commission which is chaired by the First Prime Minister of the Kyrgyz Republic to oversee climate-related interventions and mainstream climate adaptation in sector-based development. At the level of ministries, departments, and stakeholder institutions, government has also established "Climate Dialogue Platform" that coordinates information exchange, communication, and updates on climate change in the country.

The Kyrgyz's INDC prioritizes climate change adaptation for all vulnerable sectors, such as water, agriculture, energy, emergencies, human health, forests, and biodiversity (Ilyasov 2014). INDC's overarching goal is to "prevent climate change related damage and losses in the country by 2030" (GoKR 2015). Government has endorsed the framework regulatory document named as "Priorities of the Kyrgyz Republic for Climate Change Adaptation until 2017". Further, a number of sector-based action plans on climate change, such as (a) National Program on Climate Change Adaptation in Healthcare until 2015 (Box 4.1); (b) National Program on Climate Change Adaptation in the sector of Natural Disasters and Emergencies for 2015–2017; (c) National Program on Climate Change Adaptation in the sector of Forest and Biodiversity for 2015–2017, (d) National Program on Climate Change Adaptation in the sector of Water and Agriculture for 2016–2020 are operational in Kyrgyzstan[3].

Box 4.1 Mainstreaming Adaptation into Healthcare System in Kyrgyzstan

Impacts of climate change on the human health is evident. Assessment reports show that by 2100 climate- and weather-induced diseases (blood and heart diseases, lungs) might increase by 1.6–2.4% compared to 2010. The most vulnerable groups of people are children (0–1-year-old) and seniors (64–75 years old and above). The risk of infectious diseases due to climate-induced disasters is also high. Due to increase of temperature by 4.6–6.1 °C, the magnitude of infectious diseases will grow by 15.9–10.6%, respectively, by 2100 compared to 2005.

The already endorsed National Program on Adaptation to Climate Change contains 11 priorities including health protection measures to reduce the risk of extreme weather events. These include access to safe drinking water and food to reduce the risk of infectious diseases; health protection measures against climate-induced disasters (floods, debris flows, etc.) to reduce risk of accidents and trauma; diversification of training curriculum in light of climate change and capacity building for the health personnel; raising awareness and information exchange about climate change impacts and public health.

Source: MoH (2011).

[3] For a detailed overview of action plans, please refer to climate change centre resources platform, available at: http://climatechange.kg/publikatsii/

4.3.3 Tajikistan

Tajikistan is unfortunate to be the most vulnerable to climate change compared to other countries of Eurasia (WB 2009). Climate resilient development is a priority of the Government of Tajikistan, and it aims to reduce vulnerability through a comprehensive integration of adaptation measures into strategic planning and sector-based development by 2030 (GoT 2015). In 2015, the Government of Tajikistan renewed its National Development Strategy until 2030 and considered "urgent and immediate actions for climate change adaptation" as priorities. Tajikistan has endorsed its National Action Plan (NAP) for Climate Change Mitigation in 2003 and the National Strategy for Climate Change Adaptation is under formulation process.

Tajikistan is a participating country in the Pilot Programme for Climate Resilience (PPCR). The main focuses of the PPCR in the Republic of Tajikistan are on hydropower, renewable energy, agriculture, and forestry sectors and efforts are to develop adequate responses to risk reduction against natural disasters, provision of hydrometeorological services, as well as measures to raise public awareness and capacity building (GoT 2015). One of the examples to showcase Tajikistan's ambitions to maintain hydropower-based energy production is presented in the Box 4.2.

Box 4.2 Sustainable Hydropower Infrastructure as an Adaptation Measure in Tajikistan

Expert assessments show that climate change will impact the river flow regime and require all basin countries to revisit the operation procedures of existing water reservoirs considering their role in hydropower generation, agriculture, irrigation, and ecosystems. The project is to modernize the existing hydropower infrastructure in Kairakkum (northern part of Tajikistan) as a pilot to showcase climate-resilient adaptation measures both for sustainable irrigation and energy production. By the end of twenty-first century, it is forecasted that the basin of the HPP Kairakkum will witness an increase of water runoff by 20–30% (Syrdarya River) in case of "warm and humid" climate scenario and extreme decrease of the runoff in case of "hot and dry" scenario. The latter will cease the capacity of the water reservoir and reduce hydropower production. The proposed modernization lists the measures to prevent siltation of the reservoir at the catchment area and improve the reservoir facilities to receive surplus water flow. The construction of dams in the upstream part of the river will ensure the flood protection and maintain sedimentation levels. The HPP Kairakkum is used not only for power generation but provides irrigation for downstream districts.

Source: GoT (2014).

4.3.4 Turkmenistan

Adaptation to climate change is very well positioned in Turkmenistan's INDC and aims to reduce vulnerability and economic loss due to climate change by 2030 (GoTr 2015) and meet the goals of the National Strategy for Socio-economic Development by the same timeframe. The Government of Turkmenistan states that demand for water resources to maintain a climate-resilient development of key socio-economic sectors, such as agriculture, health, soil and land resources, ecosystems, and forestry would be as much as 5.5 billion m^3 by 2030 (GoTr 2015). Crop production will most probably drop by 20% during 2016–2030 and challenge country's food security. Hence, comprehensive action plans under the National Strategy on Climate Change (2012) are to be finalized. Estimates show that implementation of planned adaptation measures would cost the Government of Turkmenistan up to US$ 10.5 billion (GoTr 2015). Box 4.3 shows Turkmenistan's approach towards IWRM as one of the adaptation options in water sector, which accommodates multiple benefits, including technology transfer for sustainable agriculture and irrigation.

The Government of Turkmenistan strives to launch the *Regional UN Centre on Climate Change* in Ashgabat city. This was announced by the President of Turkmenistan during his speech at the high-level summit on sustainable development in 2012, in Rio (Rio + 20) and once again reconfirmed at UN Summit on Sustainable Development Goals and Post 2015 Agenda for Sustainable Development[4]. One of the areas of the regional centre is to foster the exchange of best practices and technology on climate change adaptation (Akopova 2015).

Box 4.3 Integrated Water Resource Management (IWRM) as an Adaptation Option in Turkmenistan

Climate warming is expected to tremendously impact water runoff and agricultural development in Turkmenistan. Forecasts show that the annual drop in water runoff might reach 5 km^3, unless adequate adaptation measures in water saving policy and technologies are introduced. Recognizing the transboundary nature of water resource management, the Government of Turkmenistan considers IWRM as one of adaptation options for water sector development, which can bring multiple benefits in terms of new technologies of water saving practices and cultivation of drought-resistant crops, institutionalization of community-based water bodies (e.g. water user associations, transboundary water councils), strengthening the legislative frameworks and regional cooperation. The estimates show that economic benefits from adaptation measures in water sector in Turkmenistan can rise up to US$ 175 million.

Source: Allaberdiyev et al. (2010).

[4] Speech of the President of Turkmenistan, 2015.

4.3.5 Uzbekistan

Uzbekistan does not have a targeted climate change policy document that provides a strategic framework for climate change adaptation actions (WB 2013). However, there are several environmental policies and programmes that cover a range of adaptation activities in the key socio-economic sectors. Water-borne diseases play a major role in Uzbekistan's health status, with more than 30% of households lacking access to quality drinking water. Therefore, WHO has launched a pilot programme in Tashkent and Syrdarya provinces with the goal of helping healthcare systems adapt to climate-induced diseases, particularly those arising through water stress (WB 2014). Other projects include rehabilitation of irrigation systems and promotion of sustainable agriculture to respond to climate risks. The Government of Uzbekistan prioritizes climate change mitigation and reduction of greenhouse gas emissions through energy efficiency and deployment of renewable energy, and progressed in providing an enabling environment for the local and foreign investors to contribute to climate change. Box 4.4 summarizes the efforts of Uzbekistan in capitalizing investments in low-carbon development and transfer of clean technologies in the energy sector.

Despite different status and officially recognized documents on climate adaptation, all countries of Central Asia consider adaptation and mitigation efforts equally (CAREC 2015). All countries of the sub-region foresee that the common socio-economic sectors to adapt to climate change should be tackled with a priority. For example, in agricultural sector, reconstruction of irrigation channels and water supply systems, application of crop diversification and rotation practices, introduction of new technologies, and use of integrated water resource management approach are considered unique for all countries of the region. Yet, the development status of each country and availability of resources undoubtedly bring a different picture of the adaptation progress. At the same time, the countries of Central Asia recognize

Box 4.4 Enabling Environment for Business Sector to Invest to Climate Change in Uzbekistan

Uzbekistan provides an enabling environment for foreign investors to achieve a strategic goal of low-carbon development and climate-resilient future. The estimated potential investment of Clean Development Mechanism (CDM) in oil, gas, and chemical sectors is around US$ 300 million. Indeed, Uzbekistan is one of the leading countries in Eurasia with 14 registered Clean Development Mechanism (CDM) projects. Until recently, the country has succeeded to attain US$ 24 million in the frames of CDM. This success has been ensured by a very well-developed institutional infrastructure and technical capacity of the Designated National Authority (DNA) and a tax exemption policy for foreign investors, who intend to implement CDM projects in Uzbekistan.

Source: Zavyalova (2015), Yunusov (2014).

the need for exchange and regular communication, and use the existing regional platforms. They admit that learning from each other is one of the best ways towards a targeted capacity building on climate change.

4.4 Current Gaps and Areas of Adjustment

All countries of Central Asia have relevant policies, national strategies, and legal frameworks to keep them on track towards sustainable development, poverty reduction, and disaster risk reduction. However, despite all efforts undertaken towards climate change adaptation and resilience development, there are several gaps and barriers, which hamper the progress. They vary from the lack of policies and targeted programmes on climate change and absence of climate change indicators (except those reflected in INDCs) in the development strategies and poverty reduction documents; inadequate institutional and technical capacities; weak coordination between the national institutions and agencies; insufficient finance and ownership of the government; lack of technologies and knowledge on smart climate solutions; lack of awareness, etc.

With the adoption of the Paris Agreement in 2015, reporting to the UNFCCC Secretariat is becoming stricter. All countries should report on commitments and pledges, stated in their INDCs through implementation of the transparency and accountability regime. This means that another challenge which the countries of Central Asia might face since 2016 is capacity to report about emission reductions and benefits from adaptation measures in an accountable and transparent way.

4.4.1 Policy, Strategy, and Legal Frameworks

There is progress on development of targeted policy documents or action plans on climate change adaptation in Central Asia. The first document was Tajikistan's National Action Plan for climate change mitigation, adopted in 2003, which foresees the roles and priorities of the country towards climate mitigation and adaptation. In 2012, Turkmenistan endorsed its Climate Change Strategy, while Kyrgyzstan adopted its Strategic Document on priorities of climate change adaptation in 2013. However, most of these documents require operational action plans, which should foresee joint interventions of line ministries and departments, dealing with water and land resources, agriculture and food security, energy, human health, emergencies, forestry, and natural resources. These efforts should also include baseline indicators and targets to track and monitor the progress. Gap also exists in the lack of institutional memory and frequent rotation of policy makers. In recent years, most of the governments of Central Asia witnessed regular changes in the heads of environment, water, and hydro-meteorological agencies, which seriously challenges the process of policy formulation and decision-making.

4.4.2 Institutional Arrangements

Most of the countries in Central Asia have a Ministry or Governmental Agency of Environment (Turkmenistan and Kyrgyzstan) or Hydro-meteorology (Tajikistan and Uzbekistan) to coordinate climate activities in their countries with an exception in Kazakhstan. Since August 2014, the Ministry of Energy of Kazakhstan became responsible agency to formulate and implement low-carbon development policy. Heads of the designated ministries and agencies are the UNFCCC national focal points.[5] At the same time, almost every ministry or agency has its own targeted institutional structure to deal with climate change in the country. For example, in Kazakhstan, there is a Climate Change Coordination Centre (C4), which supports the government during climate change negotiations and high-level events, and JSC Zhasyl Damu to lead the Emission Trading Scheme (ETS) and greenhouse gas reduction. In Tajikistan, the Climate Change Centre is under the umbrella of the State Administration on Hydro-meteorology and is responsible for climate change coordination, capacity building and liaising with the governmental institutions, and international agencies. Kyrgyzstan also has its Climate Change Centre which is responsible for most of the coordination, and effective implementation of projects, and climate policy. Yet, regular monitoring of climate-related initiatives, reporting results, and impacts of adaptation interventions are a major gap.

4.4.3 Integration and Mainstreaming

In Central Asia, key strategic documents on sustainable development, poverty reduction, and disaster risk reduction are usually formulated as national strategies and action plans for a period of 3–5 years. For example, most of the countries have national strategies and action plans for disaster risk reduction with the assigned national body to reduce risks and optimize preparedness of the country to natural and climate-induced disasters, and strengthen capacity and awareness of the population about the risks and responses. Yet, their regulations do not explicitly mention climate change scenarios and formulation of adaptive responses in the most vulnerable communities, while the institutional mandate of being a military agency (Ministry or Committee of emergency) does not allow these bodies to apply for international support to respond to climate risks. This explains a weak coordination between climate change adaptation (CCA) and disaster risk reduction (DRR) in the region. The Kyrgyz Republic progressed with integration of climate change into development (National Strategy of Sustainable Development for 2013–2017) and sector-based integration (healthcare, natural disasters and emergencies, forest and biodiversity, water, and agriculture). However, for the rest of the Central Asian countries a better integration and coordination of these programmes lacks an inter-agency coordination on revisiting existing line ministry's and agency's policies and regulations.

[5] Detailed information is at: http://maindb.unfccc.int/public/nfp.pl

4.4.4 *Finance*

Over the past years, access to climate finance has been growing in Central Asia. As of June 2015, the countries of the region have received US$ 278 million for climate change (CFU 2015). Most of the projects, which have received funding, are donor-driven, and foresee projects on low-carbon development. For example, Kazakhstan is a leading recipient of more than US$ 160 million (2009–2014) from Clean Technology Fund and Global Environment Facility. Tajikistan is the second largest recipient of climate fund (up to US$ 80 million during 2009–2014) targeted for climate resilience and adaptation. The Pilot Programme for Climate Resilience (PPCR) of the Climate Investment Fund (CIF) contributed the biggest share of the investments—US$ 78 million. Recently, the Government of Kyrgyz Republic received US$ 1.5 million grant from CIF to develop Strategic Program for Climate Resilience (CIF 2015). Undoubtedly, countries of Central Asia require international resources to achieve a climate-resilient future. However, there is a need to expand the use of domestic funds and ensure policy sustainability and national ownership. Existing allocations within the budgets mainly foresee annual expenses for maintaining business as usual activities of line ministries and departments[6]. None of them consider climate-resilient development in the long-term perspective.

4.4.5 *Implementation Capacity*

In order to ensure a smooth implementation of climate change policy and to undertake proactive measures on adaptation, it is important to possess technical capacity at the systemic, organizational, and individual levels. In Central Asia, technical expertise on climate change mostly exists within a limited number of national institutions—mostly with Hydro-meteorology, Ministries of Environmental and Nature Protection, and respective climate change centres. In Kazakhstan, it is Ministry of Energy. Yet, most of specialists possess either specific skills (meteorologists, hydrologists) or broader knowledge (environment, water management). There is a deficiency of specialists, who have analytical skills on climate change within other agencies, both at the national and sub-regional (oblast) level. It is evident that a lot of interventions on climate change are already underway in Central Asia. However, project management and coordination is also a gap, and often neglect M&E principles. Evaluation of climate performance and reporting on INDC is also a gap today. Considering Paris Agreement that foresees submission of accurate and strict

[6]For example, in Turkmenistan, the Government allocated funds to rehabilitate water infrastructure in cotton- wheat-, and rice-cultivated lands (UNECE 2013), while Kazakhstan regularly spends for agriculture and crop productivity in irrigated lands, using water saving techniques and incentives. In Kyrgyzstan and Tajikistan, the biggest share of the budget is annually allocated for disaster risk reduction and emergencies. Yet, again, they are not considered as funds for targeted activities to minimize climate-induced disasters.

reporting on GHG reduction and resilience strengthening goals, countries in Central Asia urgently need capacity to set up MRV system to produce transparent and verifiable report.

4.4.6 Technology

Most of the Central Asian countries do not have Technology Need Assessment report, which would explicitly define what technology has proven to be effective and should be applied for each sector of economy considering climate change (GoK 2010a, b). Existing practices and needs of technologies for adaptation to climate change in the region are mostly include measures on water saving techniques and sustainable agriculture (CAREC 2012). These include (1) *water and irrigation*: micro- and drip irrigation techniques, skip furrow (alternate furrow), double-sided furrow application (double flow), surge flow, laser levelling etc., while in (2) *agriculture,* the set includes switching to less water intensive crops, drought-resistant crops, and crop rotation, zero tillage, etc. Most of these technologies have been tested at the farm level. One cannot say however, that they are used daily, as some of them (micro-irrigation techniques or laser levelling) require big investments and are too expensive to use widely. Meanwhile, practical use of technologies usually requires soft elements, like establishing or maintaining the institutional base (in the case of Central Asia: scientific institutes of academy of science, laboratories, and pilot test sites) on the ground and technical capacity of other means of communication and modelling (GIS, remote sensing).

4.4.7 Decision-Making and Institutionalization

The institutional supremacy of the main governmental agencies, which deal with climate change in the region (e.g. environment, hydro-meteorology) is relatively low than that of other ministries (economy, agriculture, energy). The lack of power sometimes creates a barrier in climate change policy making and coordination with other ministries and departments at the national level. On the other hand, other institutions (economy, agriculture, energy) while recognizing a strong coherence between their sector development and climate change, do not always possess proper technical capacity. For example, in most of the institutions of Central Asia, the subject of climate change is not included in their regulations, development plans, or mandate. Even though there are a number of interventions on climate change (country-wide, local or community-based) in the countries of Central Asia, the lack of communication and coordination between the agencies creates a barrier towards planning and implementation of activities. The process of the preparation of the National Communications on climate change is the only inter-agency platform, which enable bringing together different experts and specialists from line ministries and departments. The Kyrgyz Republic, in order to minimize the barrier with inter-agency coordination, has established a "Climate Dialogue Platform" to regularly inform the ministries and departments on ongoing climate change interventions, while Tajikistan's Government has established a working group representing a diversity of national counterparts to develop its new National Strategy on climate change adaptation. However, there's a concern that these institutions and processes are ad hoc and set to ensure a participatory approach for a particular exercise or project, rather than a natural process of interactions among different counterparts.

4.4.8 Awareness, Knowledge, and Access to the Information

In Central Asia, most of the projects on climate change include activities on raising awareness and education. Target groups may vary from policy makers to the community-based recipients. Since the access to external climate finance has been growing over the past years in Central Asia, it is assumed that the level of general awareness on climate change has also increased. However, the biggest challenge is that mass campaigns do not usually foresee the involvement of media, rather counting on civil society organization and NGOs. This creates a gap in long-term strategy on awareness and information exchange and minimizes the overall impact on general population, as the main source of outreach in Central Asia is played through mass media. On the other hand, despite many targeted awareness programmes, there is still a perception that climate change is an absolutely environmental challenge rather than an issue of development and should be tackled by respective "environmental" institutions.

4.5 Readiness for Enhanced Adaptation Actions

Based on information presented above and expert judgement, this section presents the level of readiness of the countries of Central Asia towards enhanced adaptation actions. Overall, it is evident that the region has been fostering its efforts on climate policies, integration of adaptation actions into development plans and programmes, and capacity building. Countries have also received quite big investment envelope that aims enhancing actions on climate adaptation and resilience building. At the same time, it is evident that the countries of Central Asia have progressed a lot on prioritizing their socio-economic sectors, highlighting water, agriculture, and energy as top three sectors, which should fit into the national adaptation agenda. However, the biggest challenge for mainstreaming, implementation, and institutionalization of climate change in Central Asia is a poor inter-agency coordination and lack of capacity, competencies, and skills of the specialists in line ministries and agencies. Access to climate finance, although successful in getting external resources, needs to expand budgetary allocations for long-term climate resilience.

The same applies to emerging but not permanently based technologies, which could be effectively used and transferred among the countries.

The need for a participatory approach in the decision-making process is slowly but steadily recognized in Central Asia. However, tools and mechanisms for a better decision-making are still underdeveloped and to a greater extent depend on inter-agency collaboration and capacity. The level of awareness on climate change is growing mostly due to public campaigns and individual projects. Yet, they remain episodic and do not regularly involve mass media. Table 4.5 provides a summary of the current status of readiness, adequacy, and its effectiveness of adaptation actions.

4.6 Attaining Adaptation Vision in Central Asian Region

Climate change is no longer a projection for Central Asia. Increased temperature, fluctuation of atmospheric precipitation, and melting of mountain glaciers are evident and impacting key socio-economic sectors. A few critical gaps, which commonly hamper the adaptation progress across the sectors, have been identified. They include, but are not limited to the emerging but still inadequate climate policy and mainstreaming processes, institutional arrangement, implementation capacity, inadequate technology transfer and decision-making tools and processes, awareness raising. Based on information available and expert judgement, the following approaches and measures have been identified for countries to consider for enhancing adaption actions.

(a) *Promote proactive approaches* to mainstream climate change adaptation into key priority sectors, including the national development strategies and programmes on poverty reduction, line agencies action plans on water resource management and land reclamation, agriculture, food security, energy production and hydropower generation, disaster risk management and human health, protection of natural ecosystems and biodiversity, and forest management. There are emerging policies and framework documents available in Central Asia. However, a tangible and indicator-based approach is needed to revisit the existing programmes and policy frameworks, considering expected changes in population dynamics, climate scenarios, and global climate commitments by 2030.

(b) *Close a gap* between the national strategies and policies on climate change and community-based needs in adaptation. There are observations that the needs of the local population remain isolated from the policy making process, especially during formulation of adaptation measures. It is also important to mainstream climate responses not only to the national programmes and policy documents (including sector-based) but to integrate them to the local development plans, foreseeing a bottom-up approach. This can be embedded through the oblast development strategy, which is an integral part of the national development strategy in the countries of Central Asia.

Table 4.5 Readiness of Central Asia to adapt to climate change[a]

		Effectiveness in formulation and implementation		
Indicators	Adequacy	Presence/absence of instruments and practices	Level of compliance[b]	Comments
Policy, strategy, legal provision, and governance	Independent sector-based policies are emerging Lack of sector-based adaptation plans of actions	Stand-by good practices are available and need to be synchronized with the overall development strategy	*	There is a growing interest among the countries of CA to exchange good practices on policies and legal frameworks. The regional organization is keen to provide a facilitative support
Institutional arrangement	Lack of coordination Institutional arrangement is in place	Instruments and practices are present, however remain weak; BAU approach is mostly used	*	Line ministries and departments recognize the need for enhanced inter-agency coordination, but institutional bureaucracy hampers the progress
Integration and mainstreaming	Priorities for the most vulnerable sectors of socio-economic development have been set Lack of understanding and institutional capacity Lack of correlation between climate and development policy Good examples of mainstreaming in Kyrgyzstan	Examples of mainstreaming are usually project-driven, and need to be regularly monitored	*	The need for mainstreaming climate change adaptation into development policies and action plans of line agencies is becoming more prominent over the recent years. Yet, a target-based approach is still missing
Finance	Access to climate finance is increasing Yet, there is no targeted state budget for climate change adaptation	Largely absent, based on ad hoc and donor-driven demand	*	Almost all countries have access to climate investments from external resources. However, there is still a great potential to improve the access to resources internally or regionally Countries might think of preparing a Public Expenditure Review on CC—to define the existing resources and plan future expenses

Implementation capacity	Lack of technical competence and knowledge Limited competence on project management cycle, M&E, and MRV	Instruments and practices are present, however remain weak and require innovative solutions	*	Regular trainings and capacity buildings exercises are conducted to close the capacity gaps. For example, preparation of NCs is a good platform for capacity building and implementation
Technology (development, utilization, capacity)	There are emerging technologies, which have been proved effective for water and agriculture sector at the community level Weak capacity of the academia and research institutions to explore new technologies	Instruments and practices are there, but should be structured, updated, tested, and replicated	–	Efforts are needed to close the gap associated with weak scientific base, expensiveness of the service and machineries on the ground Technology Needs Assessment (TNA) can be an exercise to better understand availability and needs of adaptation technologies in CA
Decision-making tools and institutionalization	Institutional power is still weak and requires an enhanced inter-agency coordination	Some countries of Central Asia have prioritized measures for a medium-term adaptation actions Decision-making tools are not adequate; participatory approach and democracy is still weak	–	Democratic tools of decision-making process are needed
Awareness, knowledge, and access to information	Level of awareness and understanding about climate change impacts is growing Large gaps exist in mass awareness and overall impact on people	Present (campaigns, mass actions, targeted awareness exercise), but weakly perceived	+	A targeted awareness raising programme on climate change for the region is needed to coordinate the sporadic campaigns Media must take lead in raising mass awareness and have a prominent effect on population

[a]Expert perception
[b](+) Adequate; (–) Inadequate; (*) Emerging

(c) *Foster a dialogue on inter-agency coordination* and communication especially those governmental institutions dealing with the development sectors at risk (water, agriculture, energy, etc.) both horizontally (national level) and vertically (national-sub-regional/oblast-local) and ensure a regular mechanism of information sharing and knowledge exchange. The existing dialogue platforms, inter-agency working groups, and in-country climate change centres can play a coordination role and further strengthen the capacity of civil servants.
(d) *Empower national institutions* to deal with climate change interventions and finance management, and explore opportunities for expanding their duties to the regular monitoring, evaluation and reporting on climate change interventions in the country, setting the baseline and tangible indicators for new emerging projects on development and climate resilience. A regular stocktaking exercise or innovative tools will help to assess the effectiveness of the project interventions, keep the records of sensitivity/vulnerability tendency, identify gaps, and needs for further actions, and account for transparent and verifiable data records.
(e) *Improve access to climate finance* and explore opportunities of the state budget allocation, aiming at targeted climate adaptation activities. Even a small share of the state funds will increase the ownership of the national counterparts and encourage the mainstreaming of climate responses into development and sector-based planning. One of the mechanisms, supporting country and community-based interventions on adaptation could be the establishment of the National Climate Trust Fund. On the other hand, in light of Paris Agreement and pledges that developed countries have made, more opportunities on climate investments are emerging for the countries of Central Asia. For example, a comprehensive approach to see climate change as part of the complex sub-regional development policy can increase chances for resource mobilization[7].
(f) *Strengthen the capacity and skills* at institutional, technical, and individual level by developing a regional capacity building programme on climate change. Based on a capacity needs assessment, the programme will map up a set of topics and target audience (e.g. negotiation process and positioning of the countries on climate change at UNFCCC COPs; high-level climate platforms and forums for policy makers; or technology transfer for experts and community-based organizations and technical staff; monitoring and evaluation, MRV for the better implementation of the projects and transparent reporting), and formulate a joint action plan with priorities, timeframes, and resource institutions.
(g) *Strengthen the regional cooperation* among all countries of Central Asia on topics of common interests and highest concern: water, agriculture, energy, human health, and natural ecosystems. On the other hand, the countries of the region have a unique opportunity to learn from each other. For example, Uzbekistan possesses a well-recognized expertise on CDMs and NAMA and can be a knowledge transfer country to other countries of Central Asia, while Tajikistan

[7] Example is approval of Climate Adaptation and Mitigation Program for Aral Sea Basin (CAMP4ASB), which is bringing US$ 38 million from the International Development Association to the countries of Central Asia in 2016–2020.

might share lessons learned from accessing climate finance for climate adaptation and resilience. Kyrgyzstan might share lessons on establishing an adequate climate change structure on governance and mainstreaming. Kazakhstan and Turkmenistan can capacitate other countries on "green" economy and technology transfer. A regional cooperation should be seen as communication and learning tool to exchange best climate-resilient practices within the region and serve as a joint policy instrument to enable a smooth integration of the regional priorities and needs into a global development processes.

(h) *Revisit the focus of awareness raising* programmes with a targeted focus on mass media and general population. Climate change is still perceived as an absolutely "environmental" challenge rather than a threat to sustainable development in most countries of Central Asia. Many campaigns on awareness raising and climate change, which are usually implemented directly by the project team, are sporadic and do not have an overwhelming impact on general population. This can be improved by involving mass media and regular long-term awareness raising campaigns. At the same time, a regular sub-regional and in-country networking, which has been proved efficient so far, should continue.

The countries of Central Asia are unfortunate to be extremely vulnerable to climate change. In light of global warming, the risks are expected to be intensified if no adequate changes are introduced into policy, institutional arrangements, finance structure, capacity, and awareness. A "business as usual" principle is no longer effective and requires innovative approaches, proactive actions, and enhancement of a regional alliance to tackle climate change and development. Paris Agreement, which has successfully entered into force in 2016, opens new opportunities for the countries of Central Asia. The momentum should be used to showcase a regional unanimity for sound and comprehensive policies on low-carbon and climate-resilient development, and adaptation actions to be better integrated into a global climate change processes.

References

Akhmetkaliyeva S (2015) Impacts of climate change in Central Asia. Overview. http://eurasianri.org/main/impacts-of-climate-change-in-central-asia/

Akopova V (2015) Status on adaption to climate change of Turkmenistan. Personal communication

Allaberdiyev G, Atamuradova I, Nurmuhamedov R, Carman R (2010) Addressing climate change in Turkmenistan: investments required in the water and electricity sectors. UNDP Climate Change Knowledge Platform

Barbone L, Reva A, Zaidi S (2010) Tajikistan's key priorities for climate change adaptation. World Bank, Washington

CAREC (2011) Gap analysis on adaptation to climate change in Central Asia. Priorities, recommendations, practices. The Regional Environmental Centre for Central Asia, Almaty, Kazakhstan

CAREC (2012) Technology Needs Assessment for adaptation in water and agriculture sectors in Central Asia. The Regional Environmental Centre for Central Asia, Almaty, Kazakhstan

CAREC (2015) Global climate processes in Central Asia: achievements, opportunities and expectations. Background paper for the Climate Focal Points and Experts Network Meeting (5–6 November 2015, Almaty, Kazakhstan). The Regional Environmental Centre for Central Asia, Almaty, Kazakhstan

CFU (2015) Global and regional climate finance data. http://www.climatefundsupdate.org/data

CIF (2015) Data and updates on allocations for Central Asia. http://www.climateinvestmentfunds.org/cif/node/4

Giesen RH, Oerlemans J (2013) Climate-model-induced differences in the 21st century. Global and regional glacier contributions to sea-level rise. Climate dynamics. In: World Bank (2014). Turn down the heat. Confronting the new climate normal. World Bank, Washington

GoK (2010a) Technology needs assessment (TNA). Government of Kazakhstan. http://unfccc.int/ttclear/misc_/StaticFiles/gnwoerk_static/TNR_CRE/e9067c6e3b97459989b2196f12155ad5/de33bfc9b4aa487f9ca0a88210f48bc3.pdf

GoK (2010b) Technology needs assessment on adaptation in 2010. http://unfccc.int/ttclear/misc_/StaticFiles/gnwoerk_static/TNR_CRE/e9067c6e3b97459989b2196f12155ad5/de33bfc9b4aa487f9ca0a88210f48bc3.pdf

GoK (2013) Third-six National Communication under the UNFCCC. Ministry of Environmental Protection and Water Resources, Astana, Government of Kazakhstan. https://unfccc.int/files/national_reports/annex_i_natcom_/application/pdf/kaz_nc3,4,5,6_eng.pdf

GoK (2015) Intended nationally determined contributions of the Republic of Kazakhstan to the UNFCCC. http://www4.unfccc.int/submissions/INDC/Published%20Documents/Kazakhstan/1/INDC%20Kz_eng.pdf

GoKR (2009) Second National Communication under the UNFCCC. State Agency of Environmental Protection and Forestry, Bishkek, Government of the Kyrgyz Republic. http://prizmablog.com/wp-content/uploads/KR-2nd-National-Communication-to-UNFCCC-2009.pdf

GoKR (2015) Intended nationally determined contributions of the Kyrgyz Republic to the UNFCCC. http://www4.unfccc.int/submissions/INDC/Published%20Documents/Kyrgyzstan/1/Kyrgyzstan%20INDC%20_ENG_%20final.pdf

GoT (2008) Second National Communication under the UNFCCC. State Administration for Hydrometeorology, Dushanbe, Government of Tajikistan. http://unfccc.int/resource/docs/natc/tainc2.pdf

GoT (2002) First National Communication of the Republic of Tajikistan on climate change under the UNFCCC. State Administration for Hydrometeorology, Dushanbe, Government of Tajikistan. http://unfccc.int/resource/docs/natc/tainc1.pdf

GoT (2014) Third National Communication under the UNFCCC. State Administration for Hydrometeorology, Dushanbe, Government of Tajikistan. http://unfccc.int/resource/docs/natc/tjknc3_eng.pdf

GoT (2015) Intended nationally determined contributions of the Republic of Tajikistan to the UNFCCC. http://www4.unfccc.int/submissions/INDC/Published%20Documents/Tajikistan/1/INDC-TJK%20final%20ENG.pdf

GoTr (2010) Second National Communication under the UNFCCC. Ministry of Nature Protection, Ashgabat, Government of Turkmenistan. http://unfccc.int/resource/docs/natc/tkmnc2.pdf

GoTr (2015) Intended nationally determined contributions of Turkmenistan to the UNFCCC. http://www4.unfccc.int/submissions/INDC/Published%20Documents/Turkmenistan/1/INDC_Turkmenistan.pdf

GoUz (2008) Second National Communication under the UNFCCC. Uzhydromet, Tashkent, Government of Uzbekistan. http://unfccc.int/resource/docs/natc/uzbnc2e.pdf

Hijioka Y, Lin E, Pereira JJ, Corlett RT, Cui X, Insarov G, Lasco R, Lindgren E, Surjan A (2014) Regional aspects. Chapter 24. Asia. Climate change 2014: impacts, adaptation, and vulnerability. Contribution of working group II to the fifth assessment report of the intergovernmental panel on climate change. IPCC AR5 WGII, Cambridge University Press, Cambridge and New York

Ilyasov S (2014) Overview of climate change adaptation in Kyrgyzstan. Expert assessment report. Unpublished. In Russian
IPCC (2007) Impacts, adaptation and vulnerability 2007. Contribution of working group II to the fourth assessment report of the intergovernmental panel on climate change
IPCC (2014) Climate change 2014. Impacts, adaptation and vulnerability. Contribution of working group II to the fifth assessment report of the intergovernmental panel on climate change. Chapter 24. Asia
IWRM (2011) Basin plan on the integrated water resources management and water efficiency in the Aral Syr-Darya basin. Report. GIZ, Kyzylorda
Lioubimtseva E, Henebry GM (2009) Climate and environmental change in arid Central Asia: impacts, vulnerability, and adaptations. J Arid Environ 73:963–977. Elsvier, Ltd
Marzeion B, Jarosch AH, Hofer M (2012) Past and future sea-level change from the surface mass balance of glaciers. Cryosphere 6:1295–1322
MoH (2011) Programme of the health sector of the Kyrgyz Republic on climate change adaptation for 2011–2015. Ministry of Health, the Kyrgyz Republic, Bishkek
Mustaeva N (2013) Climate change and challenges to sustainable human development in Tajikistan. In: Human development in action: case studies from the sustainable human development course 2012. UNDP and CEU. ISBN: 978-92-95092-51-8
Mustaeva N, Wyes H, Mohr B, Kayumov A (2015) Tajikistan: country situation assessment. Working paper in the context of the CARIAA/PRISE initiative. CAREC, Almaty
Pawlowski IRA (2012) Climate risk management in Central Asian agriculture: a situation analysis. Zentrum fur Internationale Entwicklungs- und Umweltforschung der Justus-Liebig-Universitat Giessen. Gissen, ZEU
Sirojiddinov K (2012) Newspaper article "other problems of Nurek HPP". http://news.tj/ru/newspaper/article/mertvyi-gruz-ili-eshche-raz-o-problemakh-nurekskoi-ges (last seen in March16, 2015)
Thurman M (2011) Natural disaster risks in Central Asia: a synthesis. United Nations Development Programme, Bratislava
UNECE (2013) Promoting green innovation. Policy assessment and recommendations. Turkmenistan. http://www.unece.org/fileadmin/DAM/ceci/publications/Assessment_Turkmenistan.pdf
Verhoturov D (2015) We need to clean up Nurek Reservoir. In: Asia Plus newspaper. http://news.tj/ru/newspaper/article/nuzhno-pochistit-nurek
WB (2009) Adapting to climate change in Europe and Central Asia. World Bank, Washington. http://documents.worldbank.org/curated/en/127181468024643244/pdf/489480ESW0ECA010Box338935B01PUBLIC1.pdf
WB (2010) The CAMP4ASB Project appraisal document (PAD). http://www-wds.worldbank.org/external/default/WDSContentServer/WDSP/IB/2015/10/20/090224b083157ebd/2_0/Rendered/PDF/Central0Asia000l0Sea0Basin0Project0.pdf
WB (2013) Kazakhstan, Kyrgyz Republic, Tajikistan, Uzbekistan, and Turkmenistan: overview of climate change activities. Country note. World Bank, Washington. http://documents.worldbank.org/curated/en/2013/10/19185591/kazakhstan-overview-climate-change-activities
WB (2014) Turn down the heat. Confronting the new climate normal. World Bank, Washington. http://www-wds.worldbank.org/external/default/WDSContentServer/WDSP/IB/2014/11/20/000406484_20141120090713/Rendered/PDF/927040v20WP00O0ull0Report000English.pdf
WB (2015) http://data.worldbank.org/region/ECA
Yunusov K (2014) Why invest in Uzbekistan now? Online article. https://www.linkedin.com/pulse/20140627081140-27269359-why-should-investors-seriously-consider-investing-in-uzbekistan-now
Zavyalova L (2015) Supporting Uzbekistan in transition to a low-emission development path. Speech at the regional conference on climate change: opportunities for joint actions towards Paris 2015. Conference proceedings. Unpublished. CAREC, Almaty
Zoi Network (2017) Climate Policy: Central Asia, Eastern Europe and South Caucasus. Online product. https://zoinet.org/product/climate-policy-central-asia-eastern-europe-and-south-caucasus/

Chapter 5
Status of Climate Change Adaptation in Northeast Asian Region

Wanglin Yan, William Galloway, and Ju Youn Kang

Abstract Differences in the political systems and the economic conditions of each country in the northeast region indeed lead to diversity in terms of climate change vulnerabilities. These differences cast strong shadows on the types and scale of adaptation plans and measures started in the region. Keeping context specificity in mind, this chapter summarizes the perspective of climate change impacts, vulnerabilities of key sectors and society, and status of adaptation measures including development of national adaptation strategy and action plans in China, Japan, Korea, and Mongolia. Further it compares commonalities and gaps amongst the countries. Apart from a separate introduction for each nation, it also gives a comprehensive discussion focusing on the eight indicators described in the framework chapter. This chapter presents a comprehensive picture of the latest adaptation measures in four countries which will deepen understanding on current adaptation efforts and gaps for enhancing adaptation actions in each country.

Keywords Climate change · Adaptation · Mainstreaming · Integration · Policy · Technologies · Northeast Asia

5.1 Introduction

The Northeast Asia region[1] consists of the People's Republic of China, Japan, the Republic of Korea, Mongolia, and the Democratic People's Republic of Korea (Fig. 5.1). China, Japan, and Korea together play a towering economic role in the

[1] Due to limitation on information access about North Korea, this chapter focuses on four countries including People's Republic of China (China), Japan, the Republic of Korea (Korea), and Mongolia.

W. Yan (✉) · W. Galloway
Research Center for Climate Change Adaptation (RCCCA), Keio University, Tokyo, Japan
e-mail: yan@sfc.keio.ac.jp

J. Y. Kang
Korea Adaptation Center for Climate Change (KACCC)/Korea Environment Institute (KEI), Sejong, Republic of Korea
e-mail: jykang@kei.re.kr

M. Alam et al. (eds.), *Status of Climate Change Adaptation in Asia and the Pacific*, Springer Climate, https://doi.org/10.1007/978-3-319-99347-8_5

Fig. 5.1 Geographic location of the Northeast Asian countries

region and the world. China is defined by a vast and varied terrain. Accordingly, the impact of climate change (CC) is different depending on geographic conditions. The Tibetan Plateau experiences change in the form of river flows and melting glaciers; northwest China is threatened by deepening desertification because of reduced precipitation and higher evaporation rates; the north and northeast part of China faces water scarcity; the east and south-eastern China, especially the delta areas like the Yangtze River Delta and Zhujiang River Delta, are constantly savaged by increasingly frequent typhoons, storm surges, and long-term inundation. Since 2012, China has been suffering from frequent extreme weather conditions. As the National Development and Reform Commission (NDRC) has pointed out "…many areas in the south have experienced extremely high temperatures, with increased urban, regional and mountain floods, landslides and mudslides" (NDRC 2013). China will face ever more serious challenges as a result of CC if the government is not able to carry out appropriate adaptation measures designed explicitly to cope with current and potential climate risks (NDRC 2010).

Japan is a stratovolcano archipelago of 6852 islands. As a nation of islands, sea level rise could easily affect a significant percentage of the nation's population if temperature continues to increase. According to the IPCC Fifth Assessment Report (AR5), temperature will increase by 2.1–4.0 °C by the end of twenty-first century compared to the period of 1980–1999. If this comes to pass, the pressures incurred by sea level rise in Japan will be compounded. The average temperature in Japan has already risen by 1.15 °C during this century (MOEJ 2012). Climate change has caused varied kind of negative impacts in Japan, including reductions in agricultural production, extreme climate events and health problems including heat stroke (MOEJ 2009).

Korea is located on the Korean Peninsula, separated from China and the mainland by North Korea to its North. Climate change has similarly brought negative impacts to Korea. An increase of 1.2 °C from 1981 to 2010 has already been observed, 0.41 °C per decade. Temperature rise in the Korean peninsula is projected up to 2100. With an increased pace of annual average temperature rise, it is projected that the Korean peninsula will be 1.2 times warmer than the global average increase in temperature according to the RCP 4.5 and 8.5.

Mongolia is a landlocked country with a total area of 1,500,600 km^2. Its territory includes areas of relatively high altitudes, and harsh climatic conditions which act as significant brakes on its economic development (Dagvadorj et al. 2009). Anticipated CC will impose further limits and will almost certainly create hardship for the nation's residents. The average temperature in Mongolia has increased by 1.9 °C since 1940s. Permafrost is decreasing and mountain glaciers are melting because of CC. As an obvious consequence, the groundwater table in Mongolia has decreased, becoming a problem to the farming industry, most especially in naturally arid regions. Demands for water in the growing mining industry across the country are similarly difficult to manage, and problems are expected to be exacerbated by CC (Mayer 2015). At the same time, natural disasters have increased as a result of the particular sensitivity of Mongolia's climate, causing a significant problem for the nation and impacting strongly on its socio-economic situation. Concerted precipitation, snowfall, strong winds, sandstorms, snowstorms, hail, and flooding are common (Subbiah et al. 2013). Already some 2.4% of the livestock population dies each year because of severe weather conditions (Dagvadorj et al. 2009). Massive disasters named dzud have caused massive loss of livestock, causing direct damages to nomadic herders and their lifestyles (ERST_UNDP_NEMA 2010).

The impacts of CC are a result of a complex interaction between natural processes and human development/inhabitation at multiple scales. Problems observed in Northeast Asia are not unique in comparison to episodes taking place in every corner of the world. While the four countries share many of the same bitter experiences and negative impacts from CC, their relative positions within the realm of global negotiation on CC have created different interests for each country. China, Japan, Korea, and Mongolia each has a unique position in the international community in terms of their recognition of, and ability to manage, CC. China is the largest developing country and also the largest annual emitter of greenhouse gases (GHG) in the world. It has a very large population, and many aspire to become consumers at the same level of developed nations. At the same time, they are feeling pressure to reduce GHG emissions from the global community. Japan is the only Annex-I country to Kyoto Protocol and has focused only on mitigation until very recently. Korea has caught up with the developed world in the age of globalization and has become an industrialized nation. With its economic power, the country is eager to expand on its contribution to global issues. At the same time, it must deal with some of the negative aspects of rapid development. In 2008, it expressed a strong interest in mitigation and adaptation in combination with new economic growth in the form of a new national development paradigm focused on Low Carbon and Green Growth. The situation in Mongolia is fundamentally different. While it encompasses a vast land area, it maintains a comparatively sparse population. At the same time, its arid ecosystems make the country highly sensitive and vulnerable to any changes. Although its own GHG emissions have a marginal effect on the global climate, the country is severely affected by the consequences of CC in terms of precipitation patterns and temperatures.

5.2 Regional Adaptation Objectives

5.2.1 Adaptation Vision

Climate change issues have attracted intense attention in the four countries of the region. Governmental authorities and scientific researchers have made a concerted effort in each nation to assess the negative impacts of CC. The national visions and action plans on both mitigation of, and adaptation to, CC are summarized in Table 5.1.

In the wake of rapid urbanization and economic development, China is consciously trying to strike a balance between economic growth and the reduction of the negative impacts of CC. The government has acted proactively to enhance its ability to adapt to CC and respond to extreme weather and climate-related events across major sectors. A key goal of government-led adaptation activities was to improve understanding the local impacts of CC, as well as determining effective actions that could ameliorate or avoid negative impacts on planning and construction projects (NDRC 2013).

Japan's motivation to act on CC was subdued after several industrial countries withdrew from the Kyoto Protocol in the 2010s and negotiation for a post-Kyoto agreement stagnated in the Conference of the Parties (COP) negotiations. At the same time, long-term problems for the nation began to be felt, including a rapidly shrinking population and the ageing of people and infrastructure, especially in rural areas. These issues in themselves constitute a threat of a large-scale change for the country and require creative adaptation-oriented policy initiatives at the governmental level. The national vision for combating CC and other unprecedented events was dramatically revised after the Tohoku earthquake and tsunami in 2011. Japan has enacted a national programme called Resilience Japan that stretches across ministries and jurisdictional levels in order to avoid unprecedented risks from disasters including those induced by climate change.

Following the Framework Act on Low Carbon and Green Growth, Korea has formulated a comprehensive adaptation policy in 2010 called "the National Climate Change Adaptation Master Plan (NCCAMP)" which aims at promoting adaptation to CC nationwide. In doing so, Korea became the only country in Northeast Asia

Table 5.1 Adaptation vision of Northeast Asian countries

Country	Adaptation vision/goal/objective/approach	Source
China	Mainstreaming climate change adaptation into every process of social-economic development; enhancing adaptive capacity of key sectors, key regions, and people; and setting up a Chinese-style win-win development model of climate change proofing and sustainable development are key elements of China's adaptation vision. It has also set its goals to increase irrigation efficiency to 0.55; restore 50% of recoverable desertification land; and remarkably improvement of disaster reduction capacity Country has also named key approaches which include (a) balance global demands and domestic development needs; (b) simultaneously precede mitigation and adaptation; (c) enhance technological innovation and institutional innovation; and (d) combine of governmental leading and public participation to attain above stated goals	National Adaptation Strategy (NRDC 2013) National Climate Change Action Plan (2014–2020) (NRDC 2014)
Japan	Japan's vision is to make all the national land more resilient to large-scale natural disasters that may exert tremendous influence on the lives of the citizenry and the national economy. While it sets (a) risk avoidance; (b) reduction of negative impacts; (c) risk sharing; (d) risk acceptance; and (e) exploitation of opportunities are main goals. Government of Japan would like to (a) use diverse options, including "soft" and "hard" measures, (b) create synergies and prevent adverse socio-economic and other impacts, and (c) integrate mitigation and adaptation measures are main approaches to reach its goals	Basic Act for National Resilience (Act No. 95, 2013) National Plan for Adaptation (MoEJ 2015)
Korea	Climate change adaptation vision of Korea is to create a safe society and support green growth by adapting to climate change. It has eight goals which include (1) protect people from heat waves and air pollution (Health); (2) minimize damage through the consolidation of disaster prevention foundation (Disaster); (3) switch to adaptation based agricultural system and create a new opportunity (Agriculture); (4) improve forestry health and reduce forestry disaster (Forestry); (5) respond to sea level rise and secure stable marine resources (Coast/Fishery); (6) build secure water management system from flood and drought (Water Resources); (7) secure biodiversity through restoration and protection (Ecosystem); (8) provide foundation data and minimize uncertainty (Climate Change Monitoring and Prediction); (9) find new and promising industry for climate change adaptation (Adaptation Business/Energy); (10) enhance domestic and international communication on adaptation (Public Relations, Education, and International Cooperation) Establishment of Korea Adaptation Centre for Climate Change (KACCC) and integrated efforts on climate change adaptation cross policy sectors and jurisdictional levels appears to be main approaches to reach the above goals	National Climate Change Adaptation Master Plan (2011–2015) (MOE 2010)
Mongolia	Vision of Mongolia is to create a sustainable environment for development by promoting capacities and measures on adaptation to climate change, halting imbalances in the country's ecosystems and protecting them Its goals are to (a) enhance capacity of the country to adapt to climate change and desertification and reduce their negative impacts; and (b) build climate resilience at the community level through reducing risk, and helping adaptation. Assessment of impacts and vulnerabilities of environment and economic sectors caused by climate change, and alignment of new developments and amendments of policies and legal documents with recent climate change and the latest socio-economic development plan appear to be their main approaches	MDG-based Comprehensive National Development Strategy of Mongolia (CNDS 2008) National Action Programme on Climate Change (2010)

that legally established a position on CC action through the mainstreaming of green growth.

As one of the most vulnerable countries to CC, Mongolia has linked its CC action with Millennium Development Goals. While Japan has, until recently, paid most attention to risk avoidance through the promotion of activities that would resist change the other three Northeast Asian countries are relatively more focused on finding opportunities in their individual efforts to adapt to the challenges of CC. Even so, beyond this difference between developed and developing nations, the goals and approaches recommended in each nation are noticeably different. The Chinese government promises to meet specific numerical goals by 2020. Korea and Mongolia both declared their intent to decrease the negative impacts of CC and to build their climate resilience. Fortunately, all four countries currently view mitigation and adaptation to CC as a possible opportunity to direct social and environmental development in a way that builds resilience and long-term sustainability.

5.2.2 CCA Priority Sectors

Key areas of adaptation in this region are set out for each country in Table 5.2. The conservation and sustainable management of the environment resources is a significant area of concern for all four countries in the region although their nature differs depending on culture and geography. These include agriculture (livestock in the case of Mongolia), forestry, soil/desertification, water, coastal and marine resources, and biodiversity. Concerns about disaster prevention and risk reduction are also clearly shared by all four countries. These take place in the form of the establishment of monitoring, early warning system, and quick response schemes.

In China, key sectors, including disaster prevention and mitigation, monitoring and early warning, agriculture, water resources, coastal areas and ecosystem, as well as public health were selected in correspondence to relevant government divisions at various levels of the government (NDRC 2014, 2015).

The National Adaptation Plan (2015) of Japan offers basic strategies of the Government for adaptation to the impacts of climate change, and the basic directions of measures to implement in each sector for the next 10 years. It aims to minimize or avoid damage from the impacts of climate change, and create a secure, safe, and suitable society that can quickly recover from those impacts. The government-supported S-8 project conducted a comprehensive study on impact assessment and identification of adaptation measures focusing on agriculture, forestry, water resources, coastal zone safety, health, disaster prevention, and local community (MOEJ 2012). This study has contributed significantly to the formulation of National Adaptation Plan. In the future, it will be essential to make use of the latest scientific knowledge in order to accelerate well-integrated, broad-based initiatives at both the national and local levels (MOEJ 2010a).

The "National Climate Change Adaptation Master Plan (NCCAMP)" of Korea formulated in August 2010 for the period of 2011–2015 under the "Framework Act

Table 5.2 CCA priority sectors in Northeast Asian countries

Country	Agriculture	Rangeland/ livestock	Forestry	Soil/ desertification	Water resources	Coastal/ marine resources	Biodiversity	Health	Energy/ industry	Disaster/ infrastructure	Monitoring and early warning	Education	Community
China	X		X	X	X	X	X	X	X	X	X		
Japan	X		X		X	X		X		X			X
Korea	X		X		X	X	X	X	X	X	X	X	
Mongolia	X	X		X	X					X			

on Low Carbon, Green Growth" has identified seven sectors for CCA. These include health, disasters, agriculture, forestry, coast and fisheries, water resources and ecosystems, and three sectors including climate change monitoring and prediction, adaptation business/energy and public relations, and education, and international cooperation were selected for supporting activities. As a continuation of adaptation efforts, the Government of Korea has published its second National Climate Change Adaptation Master Plan (2016–2020) at the end of 2015. Its vision is Safe Society and National Happiness through Climate Change Adaptation. Under the mid and long-term goals for 2020 and 2035, it contains two programmes to build an adaptive foundation and three sectoral adaptation programmes.

High priority adaptation measures that could be undertaken within the authority and resources of the government were identified in Mongolia. These adaptation measures were focused on "...public awareness and education of herdsman; development of rangeland and livestock management systems grounded in pastoral practices as well as modern technology; improved forage production systems; development of modernized water supply systems for pastures; the establishment of appropriate risk management systems; strengthening of the early warning system within the national meteorological and hydrological services; development of an insurance system for livestock and crops with respect to natural disasters; improvement of the marketing system of livestock and crop products in coordination with long-term weather forecasts and market signals; improvement of the health care system both for people and animals etc." (MINC 2001).

5.3 Adaptation Trend in Northeast Asia

Progresses of activities in response to CC are summarized in Table 5.3. A clear trend, especially with regard to China and Korea, is visible when looking back at the last few decades. Momentum was generally begun with a programme of Action, followed by a national adaptation strategy, and finally the national adaptation master plan. Meanwhile, the difference in the stances to adaptation in the four countries is remarkable. The Chinese government, for instance, has expressed strong concerns on the subject. The first mention of the word adaptation can be traced back to Agenda 21, published in 1994; however, neither key areas nor specific tasks were defined until China's National Climate Change Programme was published in 2007. At that time, the key areas considered with regard to adaptation were agriculture, forestry and other ecosystems, water resources, and costal zones in China's National Climate Change Programme (CNCCP). These were extended to include seven sectors including infrastructure, agriculture, water resources, coastal zones and other marine areas, forest and other ecosystems, health, tourism, and other industries. This expansion is a meaningful step towards the mainstreaming of adaptation as they include industrial sector-based ministries, the driving forces of the Chinese economy.

Table 5.3 Adaptation trend and key milestones in NE Asian countries

Country	Adaptation activities
China	• First mentioned adaptation in China's Agenda 21 in 1994 • Defined four key areas of adaptation: agriculture, forestry, and other ecosystems, water resources, costal zones in CNCCP, 2007 • Emphasized that "climate change should be a must-factor in planning and constructing infrastructure, and the adaptive capacity must be enhanced in the key areas" in China's 13th NDP, 2010 • Enhancing capability across major sectors since 2012 • China's National Adaptation Strategy (2014–2020) (CNAS), 2013. Key areas extend to infrastructure, agriculture, water resources, costal zones and other marine areas, forest and other ecosystems, health, tourism, and other industries • China's National Action Programme on Climate Change (2014–2020), 2014
Japan	• Mentioned research project on adaptation technology in first NC in 1994 • One chapter for research on adaptation against sea level rising and agriculture in second NC in 1997 • One chapter on impacts and adaptation measures including coastal zone and agriculture in third NC in 2002, and fourth NC in 2006 • The number of pages on adaptation actions largely increased including natural disasters, water, food and agriculture, biodiversity, ecosystems, and health in the fifth NC, 2010 • Formulation of National Adaptation Plan and approved by cabinet in November 2015
Korea	• Formulated "National Comprehensive Plan on Climate Change Adaptation (2009–2030)" in 2008 • Developed the National Strategy for Green Growth in 2009, and the "Framework Act on Low Carbon, Green Growth" in 2010 • Developed the National Climate Change Master Plan (2011–2015) in 2010, and formulation process is going on for Second National Climate Change Master Plan (2016–2020)
Mongolia	• Identified key areas of adaptation: Rangeland and livestock, Water, Arable farming, soil degradation, and desertification in first NC in 2001 and NAPCC in 2000 • Clarified the government strategy for implementation of the adaptation measures in agriculture and water resource sectors in the MARCC 2009 • One section on "Adaptation measures, needs, challenges and opportunities" in second NC, 2010 and recommendations for each key area are listed • The importance of adaptation for the sustainable development of the country and the necessity of new development and amendments of policy are recognized in NAPCC 2010 • First comprehensive report offering information on the required technologies to adapt to the changing climate by "Technological Guidance for Technology Needs Assessment on Climate Change Adaptation in Mongolia" in 2013

Japan was aware of adaptation since 1994, like China. The attention of the Japanese government was focused on sea level rise and agriculture, a stance that lasted until the fourth national communication to UNFCCC in 2006. The fifth National Communication broadened the range of adaptation concerns to include natural disasters, water, food and agriculture, biodiversity, ecosystems, and health. However, no further information was given in the sixth National Communication in

2013 (Japan 1994, 1997, 2002, 2006, 2010, 2013), may suggesting that interest in the topic has slowed somewhat. However, initiation of formulation of national adaptation plan which has approved by the cabinet in 2015 suggests that importance of adaptation has moved beyond reporting to the international conventions.

Korea is one of fast movers in the field of adaptation. Adaptation first appeared as an official act of governance in Korea in 2005 with the formulation of the third Comprehensive Plan on Countermeasures to Climate Change (2005–2007). The first plan for adaptation, the "National Comprehensive Plan on Climate Change Adaptation (2009–2030)" was developed in 2008, and "Low Carbon and Green Growth" was adopted as a national strategy in 2009. Continuing the fast pace, in 2010 the "Framework Act on Low Carbon, Green Growth" was enacted and provided a legal foundation for adaptation activities. Accordingly, the National Climate Change Adaptation Master Plan (2011–2015) was established as Korea's first legally prescribed adaptation policy in October 2010 through the joint effort of 13 associated government ministries under the supervision of the Ministry of Environment (MOE 2015).

Mongolia was the first country in the region to define key areas for adaptation. These included rangeland and livestock, water, arable farming, soil degradation, and desertification (MNEM 2001; MNET 2010).

In summary, it can be said that China, Korea, and Mongolia have clearly noted the importance and opportunities of CC to economic growth and the sustainable development of the country. By comparison Japan, although an Annex-I country, has only paid attention to adaptation in recent years. The key areas in the S-8 project for instance are mostly limited to environmental conservation (MOEJ 2012). While adaptation in those sectors is important, the lack of momentum on adapting to climate change beyond that narrow range is putting Japan behind its neighbours in the region, if not the world.

5.4 Current Gaps and Areas of Adjustment

5.4.1 Policy, Strategy, Legal Provisions, and Governance

In general, legislatively speaking the topic of climate change was gradually improved from a ministerial programme to a national strategy, and finally enacted as Act. Japan was the earliest nation to enact legislation on global warming, but it was limited in scope to mitigation only—fulfilling commitments under Kyoto Protocol. Further, the legislation combating global warming was discarded in 2012 because of negative campaigning from the industrial sector. Korea is the only country in the region that established a basic law governing both mitigation and adaptation under the umbrella of Green Growth. Both China and Korea are advanced in legislating long-term strategy as well as short-term action plans. China, Korea, and Mongolia have developed national adaptation strategies and programmes dedicated

Table 5.4 List of national policies, acts, and programmes

Country	Strategies/acts/programmes
China	• Policy and legislations on CCA are not completed and harmonized • CCA has not been considered in most of laws and legislations
Japan	• The Act on Promotion of Global Warming Countermeasures was enacted in 1998 as the first law, and the Guideline of Measures to Prevent Global Warming was established in the same year and revised in 2002 • The Basic Law on Combating Global Warming was discarded, 2012 • Low Carbon City Promotion Act (Eco-city Law) (Law No. 84 of 2014) • Laws and legislation are concentrated on mitigation
Korea	• National Comprehensive Plan on Climate Change Adaptation (2009–2030), 2008 • National Strategy for Green Growth (2009–2050), 2009 • Framework Act on Low Carbon, Green Growth, 2010 • National Climate Change Adaptation Master Plan (2011–2015), 2010
Mongolia	• Sectorial and ministerial laws and legislations exist but lack of resolutions on CCA • National strategies of mitigation and adaptation have not been reflected in laws and other legal instruments

to adaptation activities, while Mongolia's programme covers both mitigation and adaptation. Table 5.4 below provides a list of national policies, acts, and programmes developed and enacted to support adaptation measures in four Northeast Asian countries..

China's National Development and Reform Commission, the Environment Protection and Resource Conservation Committee of the National People's Congress (NPC), the Law Committee of the NPC, the Legislative Affairs Office of the State Council, together with relevant departments, established a group to draft laws addressing CC in a bid to quicken the legal drafting process. They have also established a basic legislative framework that smooths this process (NDRC 2013). In the case of Korea, a comprehensive national policy specifically targeting adaptation was established—the National Climate Change Master Plan (2011–2015) is the main policy aimed at promoting national and local adaptation activities in the country. It was formulated by the Ministry of Environment in association with 13 relevant ministries as a 5-year rolling plan and consists of 87 major tasks over ten fields of practice.

Following scientific recommendations, Japan has developed and launched its first National Adaptation Action Plan in 2015 although it still is primarily top-down in its formulation process. Meanwhile, both at the national and local scale, Japan is taking leadership in response to IPCC assessment reports, prompting various adaptation measures against expected problems and risks through the Post-2015 framework for disaster risk reduction (HFA2). The Japanese Parliament enacted the Law Concerning the Promotion of Measures to Cope with Global Warming in 1998, and amended it in 2013. However, a higher legal arrangement in the Basic Law on Combating Global Warming was discarded in 2012 by Parliament because of opposition from the industrial sector. The first CC research in Mongolia was conducted under the US Country Studies Programme (USCSP) in 2000 (MNEM 2001). Assisted by the government of the Netherlands, Mongolia developed a National Action Programme on Climate Change (NAPCC) in 2000. It was approved by Khural, the Mongolian Parliament in 2011. The second national communication with UNFCCC was published in 2010 (MNET 2010).

5.4.2 *Institutional Arrangements*

Climate change is a cross-sectoral and cross-level issue in any society. A proper institutional arrangement is important not only for effective communication with international communities but also necessary when political decisions are made with regard to national strategy and policy. As a standard structure of the administrative arrangement, all four countries have an established cross-ministerial body in place. China, Japan, and Mongolia established a cross-ministerial committee, while Korea found a supporting institution to support the ministry taking responsibility for climate change adaptation. Institutional arrangements are provided in Table 5.5 including responsible agencies and supporting mechanism.

While the ministry hosts the international negotiation, and makes national policy, the cross-ministerial body is responsive to coordination and implementation of the policies within the context of the government at large. This is an important part of

Table 5.5 Institutional arrangements

Country	Responsible authority	Supporting mechanism
China	• National Development and Reform Committee (NDRC) coordinates climate change activities under the guidance of board of NLGACC and a cross-ministerial participation and leadership	• National Leading Group to Address Climate Change (NLGACC) was set up in 2007, with premier of the State Council as its head and 20 ministers as its members
Japan	• Ministry of Environment, one of the ministries, has limited authority in planning and implementing CC issues	• Special Group for Promoting Measures against Global Warming with premier of Cabinet as its leader and directors of ministries as members
Korea	• Ministry of Environment	• Korea Adaptation Centre for Climate Change (KACCC) was set up in 2009 to support overall adaptation work in Korea
Mongolia	• Ministry of Environment and Green Development (MEGD) • Climate Change Office (CCO) within the National Agency for Meteorology, Hydrology, and Environment Monitoring • Ministry of Infrastructure • No unified authority to take the responsibility of CC issues	• Inter-disciplinary and inter-sectoral National Climate Committee (NCC) led by MEGD with all related ministries and agencies as members

the puzzle because ministries often have divergent agendas and coordination is necessary.

China has the National Leading Group for Addressing Climate Change (NLGACC) while the National Development and Reform Commission (NDRC) became the administrative institution. Climate policies are assigned between relevant departments and local governments with a wide public participation through NLGACC.

In Korea, the Ministry of Environment Korea (MOEK) is the responsible authority for adaptation. To support research, policy-making and its implementation, networking and knowledge-sharing on adaptation, the Korea Adaptation Centre for Climate Change (KACCC) was established in July 2009 at the Korea Environment Institute. MOEK and KACCC work closely to support both central ministries and local governments as they formulate and implement adaptation plans.

In Japan, the Ministry of the Environment is the authority of the Japanese government responsible for managing actions in response to CC. The Bureau of Global Environment of the ministry undertakes tasks related to making and promoting policies in corporation with other authorities. The Meteorological Agency of Japan oversees observation networks as well as independent administrative institutions such as the National Institute of Environmental Studies (NIES) as they take part in scientific research. The Special Group for Promoting Measures against Global Warming with the premier of the cabinet is a place for coordination without administrative authority. The weak institutional arrangement might be one of the reasons for the failure to pass the Basic Law on Combating Global Warming in 2010. Because of this weak institutional arrangement, political and administrative movements in mainstreaming CC mitigation and adaptation must be made instead by sectorial and governmental departments.

The Government of Mongolia established the inter-disciplinary and inter-sectoral National Climate Committee (NCC), led by the Minister for Nature and the Environment, to coordinate and guide national activities and measures aimed to adapt to CC. High-level officials such as Deputy Ministers, State Secretaries, and Directors of the main Departments of all related ministries and agencies are members of the NCC. The NCC approves the country's climate policies and programmes, evaluates projects, and provides guidance for these activities. The responsible organizations for CC measures and actions are the Ministry of Nature and Environment (MNE) and the Ministry of Infrastructure (MI). The National Agency for Meteorology, Hydrology, and Environment Monitoring (NAMHEM), which is directly under the supervision of the Minister for Nature and the Environment, has been designated by the government as the lead agency charged with undertaking CC programmes and activities. This agency is responsible for development of policies and strategies, programmes of action, and the management and coordination of all activities related to CC issues in Mongolia. The NAMHEM is responsible also for research on CC, development of GHG emission inventories, national action programmes and plans, and the National Communications for the COP to the UNFCCC.

Table 5.6 Integration and mainstreaming approaches

Country	Integration and mainstreaming
China	• Adopted a range of major policy measures to mitigate and adapt to climate change through the 5-year plan, especially during the 11th (2006–2010) and the 12th (2011–2015)
Japan	• Most of CCA measures can be achieved in conventional programmes • Though not directly initiated as CCA disaster prevention and risk reduction including climate extremes are mostly covered by Resilience Japan Programme • Lack of mechanisms to assess what is adaptation or not from conventional programmes
Korea	• Develop NCCAMP (2011–2015) and its sectoral implementation plan in association with 13 ministries
Mongolia	• Embedded climate change action into MDG-based Comprehensive National Development Strategy (2008) • Weak coordination of sectoral actions and responsibilities are not clearly defined between sectors

5.4.3 Integration and Mainstreaming

Each of the four countries understands that adaptation needs to be grounded in development planning. The integration and mainstreaming of adaptation requires proactive activity from the public sector and cannot be managed only by the government. Failing to integrate development planning leads to the risk that adaptation planning remains confined to the pages of government documents. In this regard, there are noticeable differences in the approach of the four countries towards mainstreaming adaptation. Table 5.6 below provides a summary of integration and mainstreaming approaches of four Northeast Asian countries.

China has realized the importance of adaptation and included its consideration in the 11th and 12th 5-year plans (2006–2010 and 2011–2015, respectively). The government has additionally published an annual report on China's Policies and Action for Addressing CC. In response to the national adaptation plan, provinces including Anhui and Fujian have published their own adaptation plans.

Korea adopted their "Low Carbon and Green Growth" plan as a national strategy in 2009. Adaptation plays an important role in its enforcement ordinance and 5-year plans. Moreover, according to NCCAMP (2011–2015), 13 ministries collaborated to develop a sectoral implementation plan. It listed both existing activities which can be improved with regard to adaptation, as well as new activities designed only for adaptation. It is expected to enhance resilience within each sector, and simultaneously to promote inter-disciplinary adaptation across sectors.

In Mongolia, the NAPCC will be implemented through the integration of various existing national and sectorial action programmes, plans, and policy documents. Representative programmes include the Concept of National Security of Mongolia, State Great Khural (SGK); the National Development Concept of Mongolia, SGK; the National Concept of Ecology, SGK; the National Plan of Protected Areas, SGK; the Mongolian Action Programme for the twenty-first century (MAP-21); and the National Plan of Action to Combat Desertification in Mongolia, GoM.

In the case of Japan, while its National Adaptation Action Plan is primarily organized around a top-down concept of implementation, there is at the same time a more bottom-up system in place. After decades, if not centuries of responding to disasters, man-made and natural, the country has accumulated a body of rich experience and knowledge that is useful. This background aids the country as it explores bottom-up approaches to mitigation, and to adaptation to CC. Importantly, an integrated concept of "hard" and "soft" measures was proposed in Japan to manage adaptation (MOEJ 2010a). "Soft" measures look to non-structural, non-physical, and institutional aspects. "Hard" measures refer to structural components such as physical infrastructure, equipment, and facilities. The active combination of "hard" and "soft" aspects is aimed at activating the separate strengths and advantages of both the government and the public in participatory responses to CC adaptation.

5.4.4 Finance

It is difficult to account for direct investment with regard to adaptation and development. Most adaptation activities are implemented through conventional construction projects. Nevertheless, China and Korea have successfully earmarked financial support for adaptation programmes. According to reports, significant financial resources have been allocated in the six priority sectors of adaptation in China. For example, in 2012, the Ministry of Civil Affairs, together with the Ministry of Finance, allocated RMB 11.6 billion in natural disaster relief funds, which helped victims carry out rehabilitation and reconstruction work, and contributed to the security of their basic livelihood (NRDC 2013). The China Meteorological Administration issued China's CC Monitoring Bulletin in 2011. The administration also pushed for a general survey on climate disasters and risks, and assisted local governments in formulating climate disaster prevention plans. Financial assistance was also given in the fields of agriculture, water resources, coastal ecosystem conservation, and public health. In Korea, adaptation projects are being implemented in various sectors at both national and local levels. MOEK mostly operates the budget for research, policy-making, networking, and knowledge-sharing on adaptation, while each ministry and local government operates the budget for its own adaptation projects. To support local policy-making, a matching fund was provided to establish the local implementation plans of the NCCAMP (2011–2015) in 16 provinces and mega-cities in 2010–2011.

In the case of Mongolia, foreign financial resources will play a crucial role in the implementation of the NAPCC because the national banking system is weak and Mongolia's private companies have very small financial reserves. Therefore, activities to expand existing cooperation with international financial sources should be undertaken at the national and sectoral level to secure financial and technological support. Funding of adaptation activities is now a major theme of international climate negotiations and domestic climate policy. In this regard, Mongolia's financial support for CCA relies on: (1) State and local funds prepared for environmental

Table 5.7 Financial support for climate change actions

Country	Financial support
China	• In 2012, the Ministry of Civil Affairs, together with the Ministry of Finance, allocated RMB 11.6 billion in natural disaster relief, rehabilitation, and reconstruction, as well as safeguarded the security of their basic livelihood • The central government allocated nearly RMB 850 million to support undertakings such as the restoration and remediation of local coastal waters and the coastal zones, the ecological restoration of oceanic islands, and the protection of freshwater resources • The Ministry of Water Resources compiled or implemented the relevant plans, programmes with RMB 35.21 billion in soil and water conservation • Lack of investment mechanisms for private and public finance
Japan	• Totally US$ 16.0 billions were allocated for mitigation and adaptation in developing countries during 2013–2015 • Collect US$ 100.0 millions annually by joining the power of public and private sectors till 2020 • Lack of mechanism to involve private sectors in domestic adaptation business
Korea	• The budget was allocated to MOEK for adaptation theme as a whole and to each ministry and local government on its own adaptation project • A matching fund was provided to establish the local implementation plan of NCCAMP (2011–2015) to 16 provinces and mega-cities in 2010–2011
Mongolia	• Lack of financial resources for initial investments stands to limit the implementation of the identified measures • No existing legal regulation to promote the economic interests of private companies, state owned facilities, and communities in the area of CC • (1) State and local funds prepared to environmental issues; (2) Financial support from donor countries (UN Agencies, International organization); (3) Local and international investment and credits; (4) Donations from governmental and NPO, private sector, and regular citizen

issues; (2) Financial support from donor countries (UN Agencies, International organization); (3) Local and international investment and credits; and (4) Donations from governmental and NPO, private sector, and regular citizens.

By contrast, Japan allocated financial resources through conventional budgetary systems through several individual sectors. This amounted to $16 billion for mitigation and adaptation during 2013–2015, in which US$ 13.0 billion was public and US$ 3.0 billion from the private sector (Japan 2010, 2013).

As shown in Table 5.7, financial resources for mitigation and adaptation are bundled together, without applying a specific mechanism for the support of adaptation activities. A common issue for the four countries is how to get the involvement of private enterprises in adaptation. Especially, it is important to increase interest and awareness of the subject even when a business or sector is not affected directly through a chance disaster. In response to this concern, the Japanese Ministry of Economy, Trade, and Industry developed a special programme to encourage private companies to develop adaptation businesses from 2012. Five to seven projects were granted financial aid each year since the programme began, most of them focused on Base of Pyramid (BoP) business concepts. The opportunity to develop an adaptation industry, focusing on infrastructure and disaster prevention, remains undeveloped in the international market. This can be perceived as a clear and challenging gap that needs to be targeted actively in all four Northeast Asian nations.

5.4.5 *Implementation Capacity*

Each country in this region, except for Japan, has established both political and institutional arrangements for CCA. China so far has defined adaptation principles and conducted pilot projects as low as the provincial level. From the perspective of governmental policy and national planning, China is suffering from the consequences of its previous short-sighted adaptation planning. At the same time, the country is suffering from natural and man-made disasters because of massive development that to date has left no room for adaptive planning that might cope with unknown climate events. Considering the difference between mitigation and adaptation planning, it is hard to measure the exact impact of adaptation when it is implemented through a top-down approach. In Korea, the MOEK established KACCC to support national and local adaptation activities. It regularly provides training for national and local government officials and formed an advisory group to support adaptation projects carried out by local governments. It also held seminars and workshops to support various adaptation actors including business, civil society, and academia.

In the case of Mongolia, one of the central difficulties faced by the nation with regard to climate change adaptation is a lack of implementation capacity. The lack is not a matter of will or intention but rather of limited human resources as well as a lagging information and social infrastructure. People expect very much of local governments in terms of relief and recovery from climate disasters; however, local governments do not generally have enough resources, have poorly structured personnel management, and a low degree of autonomy. It is therefore important for Mongolia to strengthen and reform local government bodies. This means building a result-oriented management mechanism, and strengthening governance skills amongst local political leaders. The ability to undertake strategic management, to think critically and learn new things, to mobilize and integrate governance resources, and the ability to coordinate, negotiate, and resolve conflicts in essential (Chou 2007).

Table 5.8 below provides a summary of implementation capacity of four countries in Northeast Asia. Capacity building is a long-term issue in any context. Capacity building in response to CC is particularly time consuming, and adaptation to CC is even more so. Japan has a rich experience in disaster management and an impressive advantage in human resources which could speed up the implementation of its recently developed National Adaptation Plan. Further, these could be taken advantage of if national policy were aligned with the needs of CCA. To date that alignment is not yet efficient.

Table 5.8 Implementation capacity

Country	Implementation capacity
China	• NAS defined the principles and approaches of implementation: (1) key areas of priority, (2) proactive adaptation; (3) rational adaptation; (4) collaboration horizontally and vertically; (5) public involvement • Pilot studies at provincial level have been conducted such as Anhui, Fujian • Lack of clear definition of the principles and approaches; lack of operational options for governmental and industrial sectors
Japan	• Majority of short-term adaptation actions can be bundled with disaster risk reduction. Long-term actions are combined with mitigation through low-carbon society projects • International assistance for adaptation in developing countries through ODA • National adaptation approaches were recommended by the special committee of MOE
Korea	• KACCC is supporting various adaptation actors with training, advisory group, and events
Mongolia	• No coordination of sectoral actions and shared responsibilities are not clearly distinguished between sectors: (1) Mandates are not identified, or are not clearly allocated to a department; (2) there are overlaps in competence, leading to interdepartmental conflicts; (3) the management of the sector is therefore far from optimal

5.4.6 Technology

Technology can play many roles in relation to adaptation. One of the more essential roles to play is found in monitoring and early warning technology. All four of the Northeast Asian nations reorganize the importance and the need to continually enhance and improve early warning technology, including tools to analyse and communicate the data in order to make it palatable and useful for all potential users. China has recently established a functioning platform for public meteorological services, improved its climate observation network, and heightened the timeliness of meteorological services. They also improved the ability to forecast and issue warnings in the case of meteorological disasters and similarly improved the ability to respond to climate and weather related disasters. Similarly, Japan has developed one of the most advanced climate monitoring systems in the world. Specifically, the ministry of Internal Affairs and Communications developed a series of early warning systems called J-Alert and L-Alert. The former was begun in 2007 as a nationwide early warning system that could speed evacuation times in the case of disaster. The L-Alert system gathers and broadcasts local information using several presentation platforms including television and the internet. It is aimed more directly at local governments and residents. Currently, the Ministry of Internal Affairs and Communications is working on the promotion and enhancement of the L-Alert system so that it will be more effective as a communication tool.

Technology is equally important as a tool to be used in improving the accuracy of climate risk and impact assessment. In order to support local government's implementation planning of NCCAMP (2011–2015), Korea developed vulnerability maps at the small (provincial and municipal) scale and built an online service platform called Vulnerability Assessment Tools to Build Climate Change Adaptation Plan (VESTAP). This helps local governments to understand the specific impacts of climate change in their own locality and highlights their vulnerability as well as their ability to respond to them. The Japanese "S-8 project" was designed to reassess risks with respect to IPCC scenarios, making use of the latest data and information. Going forward, adaptation measures will need to be refined based on new information. As risk continues to change, so must the planned response.

A third major use of technology is the invention and application of technology purposefully directed at CC adaptation. This kind of technological effort can become a trigger for transformational adaptation. Adaptation research is yet needed in order to understand what society needs to adapt to in concrete terms, to identify available options, and to understand how adaptation can be effectively promoted and implemented. By way of example, China's Ministry of Agriculture has placed more than 420,000 germplasms in long-term storage, making it the second largest collection in the world (NDRC 2010). Water saving technology is also being promoted by the national government in the form of demonstration projects. Korea actively promoted green technology that additionally embraced adaptation technology and established the Green Technology Centre (GTC) in 2012 to support policy-making and cooperation in the realm of green technology. Following a similar path, in Mongolia the Climate Change Coordination Office of the Ministry of Environment and Green Development conducted a technology needs assessment on CCA (MEGD 2013). The resulting "Technology Needs Assessment Report" describes key adaptation technologies required in priority sectors for the country. These include a system of wheat crop intensification; a vegetable production system with drip irrigation and mulch, potato seed production system; seasonal prediction and livestock planning; early warning systems; improved livestock through breeding and animal disease management; and finally, sustainable pasture management. It also identified barriers in implementation and proposed actions to be taken in cooperation with internal and external groups. Table 5.9 below provides a summary of application of different technologies for climate change adaptation and gaps.

5.4.7 Decision-Making Tool and Institutionalization

Standardization of adaptation methods and actions is not yet in place in any of the four countries. For instance, there is no standard methodology to monitor, assess, and provide feedback on the performance of adaptation activities. In part this lack of clarity is because adaptation is generally considered to be an option in conventional planning and within institutions. This is slowly improving but it remains a significant gap and a hurdle that needs to be overcome. China was the "first major developing economy to issue an action plan" (National Development and Reform Commission 2007) in the region, when they released their CC programme in 2007. The following year the government released a white paper called China's Policies

Table 5.9 Use of different technologies for climate change adaptation and gaps

Country	Technology
China	• Conventional technologies are less of applicability to CCA • Infrastructures are not constructed under the consideration of CCA
Japan	• Ministry of Environment organized research project called S-8, assessed the impacts of climate change, loss and damage by climatic disasters, and recommended approaches for implementing adaptation measures • Developed common information platform, database systems, and monitoring networks
Korea	• Developed the vulnerability maps at a local scale and VESTAP for local governments • Promoted green technology and established Green Technology Centre (GTC)
Mongolia	• Observations and research are conducted at several national institutes separately • Needs to establish a nationally coordinated research programme on climate change to ensure the improved coordination of activities designed to monitor and analyse climate change • The systematic observation network is inadequate in the vast territory

and Actions for Addressing CC (Policies and Actions); and in 2009, they followed up with a Progress Report on those policies and actions. Notably, they devote sections to the "present and imminent task" of adaptation, as contrasted with the "long and arduous challenge" of mitigation. An analysis by the World Resources Institute calls the plan "impressive", and generally "large in scope and scale". Broadly speaking, the plan stresses the "need for adaptation of human and natural systems without hindering economic development. There is also a strong focus on national level policy/legislative approaches designed to enhance China's overall adaptive capacity". However, the plan "lacks specific targets and action-steps for realizing" their goals (Farber 2011). This too is a gap that needs overcoming in all the countries in the region.

Japan's Ministry of Environment (MOEJ) initiated the S-4 research Project for Comprehensive Projection of CC Impacts, a Global Environment Research Fund Strategic R&D Area Project. The project was completed in FY2009. That project revealed that large climate-induced impacts will appear in key areas such as water resources, forests, agriculture, coastal zones, and human health, and that there are especially vulnerable regions for each. Building on this work the S-8 project, called the Comprehensive Study on Impact Assessment and Adaptation for CC, began in 2010 with the goal of formulating adaptation measures and feasibility assessments with regard to realizing a safe and secure CC adaptive society (Mimura et al. 2014). The objectives of the project include development of an advanced impact/adaptation assessment model (specifically a bottom-up model); to obtain a more detailed understanding of the physical and economic impacts, as well as an estimation of the effects of implementing various adaptation measures. In addition, it aimed to develop monitoring methods at the prefectural and municipal levels in order to understand issues at that level of detail. Additionally, the project also analysed the order of priority and cost-effectiveness of implementing adaptation measures elsewhere in the Asia-Pacific region. This section included developing indexes for the assessment of vulnerability, impacts, and adaptation effects as applicable to developing countries. Going forward these are to be developed and standardized, in part filling the gap in capability outlined above.

Korea has likewise developed and made available various decision-making tools for adaptation stakeholders: (1) Model of Integrated Climate Change Impact and Vulnerability Evaluation (MOTIVE); (2) Vulnerability Assessment Tools to Build Climate Change Adaptation Plan (VESTAP); and (3) Climate Change Risk Assessment Tool for Business (CRAS). NCCAMP (2011–2015) did not include a monitoring and evaluation scheme in its output. However, MOEK together with KACCC monitored progress of local government's policy-making on adaptation, and plans to include a monitoring and evaluation scheme in the next adaptation master plan.

The Mongolian government, both at the national and local level, is increasingly paying attention to the strengthening of the state's capacity to cope with disasters. "All 21 provinces of the country have an Emergency Management Unit (a department or a division) responsible for disaster prevention, recovery, reduction of risks and vulnerabilities, and assessment of the past emergencies and disasters within the boundaries of the respective province. These units operate as implementing agencies beside the provincial Governor's Office and are fully financed from the state budget" (Amgalanbayar 2011). Meanwhile, information-based decision support tools were developed in order to reduce risks in the wake of a changing climate in Mongolia. For example, the decision support system for agro-technology transfer (DSSAT) 4.0 model was used to assess future trends of spring wheat yield (Addison and Brown 2014). Similarly, the "National Livestock Register" system was developed and implemented in 21 provinces and 327 soems (counties) and more than 40% of livestock were registered in pilot project regions. On a more technological level Japan's Keio University, working in cooperation with the National Development Institute of Mongolia developed a mobile phone based short message service (SMS), delivering more accurate weather information to herders.

5.4.8 Awareness and Access to Information

It can be said that public and governmental concern in the four countries is still mainly attuned to mitigation issues, while adaptation is generally treated as a supplement in government reports, national act, master plan, and programmes. There is a significant attention gap that affects policy as well as action taken on the ground. Table 5.10 below provides a summary of gaps in awareness and information availability in four Northeast Asian countries.

China's National CC Policies and Strategies were mostly focused on the reduction of CO2 emissions. Adaptation is relatively a new topic for most Chinese people and in this way, despite some progress, actions to offset CC have initially been slow and poorly coordinated. A lack of scientific research and an insufficient awareness about the need for adaptation leads unsurprisingly to a limited belief in the need for

Table 5.10 General gaps in awareness and information availability

Country	Awareness and information availability
China	• Strong evidence of top-down approach and lack of mechanism for public participation • Necessity to include CCA into primary education, high education, and professional training • Limit of information available for public
Japan	• Combination of hard and soft approach is recommended • Open data policy is actively discussed • Public awareness is relatively high on global warming • Main concerns on adaptation are for developing countries
Korea	• Efforts on information spreading and awareness improving • Promotion workshops, events, and projects are organized
Mongolia	• Early warning bulletins on winter forecasts are being received in September and on summer forecasts are received in March every year • Ministry of Food and Agriculture receives monthly, a 10-days and a weekly forecast product from NAMEM. There is a lack of capacity in MOFA to translate the forecast products from NEMA into potential impacts and provide tailor made information for herders • Keio University and National Development Institute of Mongolia developed SMS system to deliver weather and forage information weekly

adaptation. However, typhoons, thunderstorms, droughts, and urban flooding are on the rise, and these experiences have begun to change perception. At the same time, weather forecasting information and climate prediction are improving quickly. That notwithstanding, access to real-time monitoring data as well as long-term databases remain limited. Another issue is that the knowledge accumulated in individual experts and governmental offices are seldom delivered to the public in an understandable way. New media and social networks may change that situation quickly if the 2015 video document on PM2.5 titled "Under the Dome" by a former CCTV reporter, Ms. Cai Jing can be seen as a sign of future trends (Chai 2015).

In 2006, the Japanese Government launched the Data Integration and Analysis System (DIAS) project[2], one of five National Key Technologies, as part of the Earth Observation and Ocean Exploration System in 2006. One of the main missions of DIAS is to coordinate cutting-edge information science and technology amongst the various research fields addressing terrestrial environments. It is also aimed at building a data infrastructure that can integrate earth observation data, numerical model outputs, and socio-economic data effectively. With these tools at hand, it is hoped that the final output will be the creation of new knowledge that will give access to better solutions to environment problems while also generating socio-economic benefits.

Korea has made various efforts to raise awareness of adaptation and spread achievement of national policies. Newsletter on adaptation in Korean and English has been distributed regularly. Various seminars and events on latest adaptation issues have been held. "Climate Change Adaptation Supporters" which developed

[2] http://www.editoria.u-tokyo.ac.jp/projects/dias/?locale=en

educational materials and taught students at elementary schools were operated to improve general public's understanding of adaptation. The Government of Korea also has created a clearing house to collect and service adaptation-related information and data located in various institutions both inside and outside of the nation. The system aims for the greatest possible convenience for users looking for information about CCA.

The Government of Mongolia is working to increase public awareness of CC and its impact and is forming partnerships with industries and NGOs to enhance the effectiveness of measures for mitigation and adaptation (MNET 2010). Moreover, on the point of coordination, the CC Coordination Office of Mongolia is working closely with other government and research groups. Along with the constituent parts of the Mongolia government, they regularly hold informal discussion meeting with stakeholders once or twice a month. Workshops and seminars for related projects, such as CC risk assessment, form the remainder of the Mongolian effort to deal with CCA. Other projects include the CCA project, as well as a project aimed to measure CC and human health. These are both organized by the CC coordination office.

5.5 Readiness for Enhanced Adaptation Actions

Pelling (Pelling 2011) frames climate adaptation in three stages: resilience, transition, and transformation. Resilience means that a group or area absorbs external changes or shocks. Transition means the system under stress is adjusted incrementally. Transformation, on the other hand, indicates a shift in current systems to an entirely new and discrete paradigm. This framework is usefully applied in order to assess the readiness and preparedness of the countries in the northeast region in terms of their adaptation activities.

Regarding the first phase, all four countries have assessed the negative impacts that could be caused by changing climate, and the necessity for adaptation is becoming the national consensus. Global warming may raise sea level up to 100 mm in 2040 in Yellow sea, East Sea, and South Sea of China (NDRC 2014). North China may become drier, and the frequency and intensity of climate disasters could increase drastically. The Ministry of the Environment in Japan published their latest report in 2014 as part of their S-8 project and unveiled potentially severe domestic impacts (MOEJ 2010b). Loss and damage induced by climate disaster, without any adaptation, would be as much as 2–3 times the current costs due to already accounted for disaster events for the nation (i.e., earthquakes, typhoon). Willing or not, we are entering the age of CC. In addition to mitigation, adaptation must be a top priority of governmental policy.

Regarding to the second phase, each country has established an emergency unit to respond to natural and human disasters. These play a central role in relief, rescue, and recovery; however, the establishment of mechanisms to avoid or reduce unprecedented and unplanned for risk is still not well developed in China Korea or Mongolia.

As for the third phase, the four countries commonly have understood the importance of adaptation for sustainable development. As evidence, the Chinese Government requires all ministries and provinces to take CCA into account in the 13th 5-year development plan. Meanwhile, Korea has set up a national adaptation master plan that works at both the federal and local level. Mongolia has included CCA in their Millennium Development Goals. Similarly, the programme of Resilience Japan was enacted as an initiative of the Ministry of Land and Transportation, and a large budget was provided to assess the vulnerability of infrastructure, to revitalize degraded facilities, and to enhance capacity of costal zones against high tides and tsunami disasters. Japan is also devoting a lot of their attention to the "soft" aspects of capacity development at the community level—for example, the Ministry of Environment compiled possible soft measures under the title of "Wise Adaptation" (MOEJ 2008). As a method for gathering local knowledge that can be implemented at scale, it could be a good reference for the other three countries.

That said, generally speaking in the northeast Asian region there is a deficit of success in the shift from simplistic short-term responses to disaster, which amounts to absorbing the damage, towards a long-term (and more complicated) response. Namely the transformation of national goals. This can be recognized in the lack of established legislation and indicators to track the implementation of activities; the lack of specific financial resources and tools to enhance adaptation actions; the lack of unified measurement system to judge the orientation or category of adaptation activities; and a lack of unified standard to measure the performance of climate programmes undertaken at the scale of communities.

Fundamentally, though the vision of the four countries with regard to CCA appear similar on paper, large differences in the contents and implementation capacity need to be understood. Japan, historically the first so-called developed nation in the region, sees adaptation as a way to maintain their current pattern of socio-economic development, which is resilient to change. Korea, as a relative newcomer to the developed country club, sees adaptation as a way to bolster the economy and by 2050, they expect to be able to sustainably maintain their advanced status. China as a still-emerging economy is focused on reaching its goal of building a country with a significant middle-class by 2050. Mongolia on the other hand is the least developed country in the region, and their adaptation plans are aimed at reaching millennium developments goals. Unsurprisingly, each country has developed a vision according to its own baseline, a function of the capacity of their social context as well as a summation of their development achievements to date. With such a divergent ambition, it is difficult to do a direct comparison between the countries in terms of Pellings' terms outlined above; "transitional", "incremental", and "transformative". In the same way, the assumptions we can make about capacity and opportunity differ for each nation. To illustrate the difficulty, Japan, with its well-developed infrastructure, has a higher capacity to prevent disasters. With its large population, territory, and economy, China's ability to absorb shocks might be higher. On the other hand, a small incremental shift within such a large context may lead to a very large transformative change.

In the same way, if we look at the governmental policies, strategies, programmes, and plans for adaptation from the narrow perspective of environmental and climate response management, Japan lags behind the other three countries. Some Japanese believe adaptation is an issue for developing countries, and not for Japan. And yet it is the most advanced nation in the region, and perhaps even the world, about disaster management and risk reduction. "Resilient Japan" might be thought of as a kind of incremental action considering their starting point. However, those same actions would be transformational and far-reaching in developing countries. Therefore, consideration of the objectives and approaches to CCA should be context dependent, and assessment of the readiness for adaptation should be relative to the baseline.

This does not mean the four countries have nothing to share. To the contrary, the transition from a minimal state of readiness and preparedness to a more capable position can be understood as a direct measurement of the process of capacity building (for implementation). Japan has made great contribution to the world by hosting international conferences on disaster prevention and risk reduction. Assisting developing countries to enhance adaptation capacity is listed at the top of its adaptation policies. China, Korea, and Mongolia have also recognized the importance of international cooperation. In this light, China is trying to increase both domestic and international cooperation. More organizations and institutions within China are contributing to the global knowledge base on CC impacts and adaptation, especially as it occurs in developing countries. China is at the same time looking for support amongst Asian counties that share similar geological features and have the advantage of being neighbours. As part of this process, the NDRC developed a national strategy for CCA, based on an assessment of CC impacts in the areas of economic and social development, lays out "clear guidelines and principles for CCA, as well as proposed adaptive goals, major tasks, regional patterns and safeguarding measures" (NDRC 2013). A joint China-UK-Swiss project named Adapting to CC in China (ACCC) was established in order to develop accurate regional/national climate models. An important aim of the project is to integrate physical and social sciences in order to improve impact/vulnerability/risk assessments. Fields of interest for this project include Agriculture, Water resources, Extreme weather/disaster risk reduction, Public health, and Grasslands-livestock.

5.6 Attaining Adaptation Vision

There are large differences between the four countries, not only in terms of their physical scale and population, but also their political stance, economic capacity, citizen awareness, and scientific achievements. With regard to the international standard of CO2 emissions reduction, a long list of comparable research and reports on mitigation actions in this region exists. These looks to energy policy, green technology, financial support, international cooperation, and so on. Although a lot of work has been accomplished, there is still a significantly large gap between reality and the ambitions of each of the nations in terms of CCA. Much of it is intrinsic to the problem. Unlike mitigation, there is insufficient data available on adaptation, and few established indicators that can work across such a broad range of differences. Still, it is a challenge to align the four countries and assess achievements in the field of adaptation. Searching for methods to compare inputs, outputs, and outcomes in CCA is itself a major and ongoing global challenge.

Realizing an adaptation vision can only be achieved along with the progress of social-economic development. Adaptation plans in the northeast region of Asia therefore need to be developed with care and consideration of the local situation. Whatever the methods used to reach a goal, they must synchronize with national plans. The authors believe that the adaptation of a society can only be incremental although some people would like to take the opportunity to leap-frog development. Countries in Northeast Asia are undergoing different levels of development both economically and environmentally as illustrated in Fig. 5.2. Climate change action, especially adaptation shall be the culmination of synthesis of all stakeholders at all levels in every part of society.

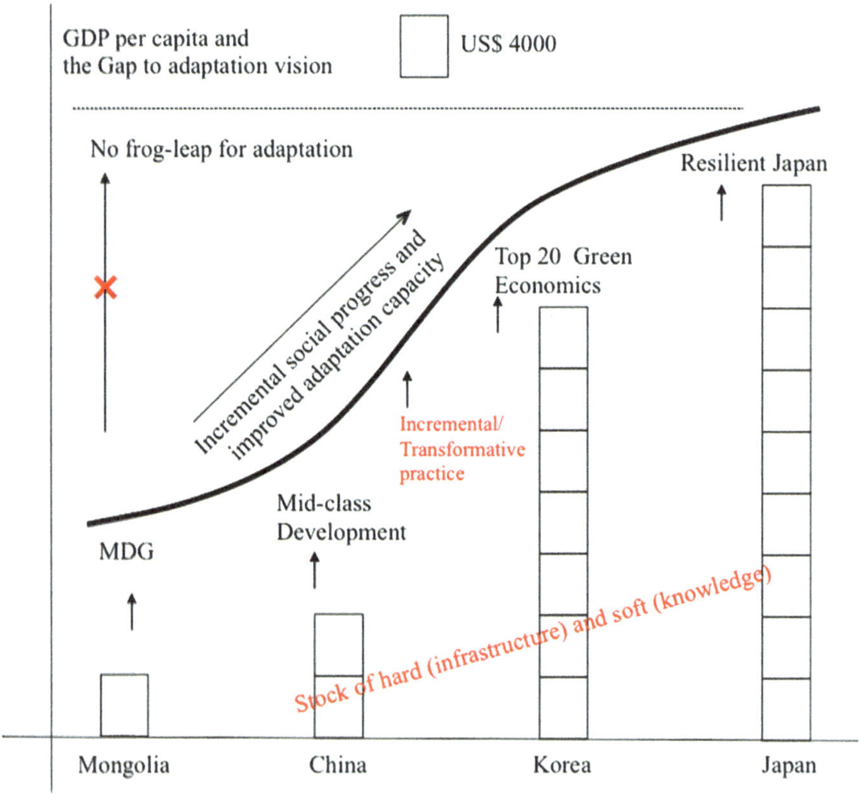

Fig. 5.2 Summary of gaps behind the adaptation visions in Northeast Asia

We cannot substitute the government-led approach that is normal in China with the "soft" approach proposed in Japan. In the same way, we cannot transfer the adaptive technology in agriculture developed in Korea to Mongolia very easily. The process must necessarily take the form of a kind of co-evolutionary learning, based on the social-ecological systems already rooted in each society.

Japan has already positioned itself as a front-runner on climate adaptation planning in comparison with other countries. The National Adaptation Plan of Japan is likely to have an obvious instructive and illustrative role for other counties. Conversely, China is a late starter with regard to CCA and Mongolia, while concerned with issues, has very limited resources. Korea aims to be amongst the top 20 green economies by 2050. Its adaptation actions are started not only at the level of the central government, but are also at local governments. Attentions should be paid in Korea on the continuity of the policies in the medium- and long-term perspective. Mongolia, Where the people experiences temperature change and fast desertification, is more obviously suffering than the other nations in the region and is extremely sensitive to CC.

The current situation in Northeast Asia shows a lack of corporation between the nations, not only in terms of policy but also information sharing. As close neighbours, China, Japan, Korea, and Mongolia collectively have access to a great deal of resources—physical, institutional, and intellectual. China, Japan, and Korea already have an environmental dialogue mechanism at governmental level, the Tripartite Environmental Ministers Meeting (TEMM). These four nations could easily benefit from a shared vision of cooperation on climatic issues and steps therefore should be taken to achieve that goal. However, for one reason or another, this opportunity has not been fully made use of for CC. It is perhaps because the countries have entirely different positions at the international negotiation. That said, ongoing efforts in the region provide an excellent platform for the countries to work together to tackle CC as a group.

References

Addison J, Brown C (2014) A multi-scaled analysis of the effect of climate, commodity prices, and risk on the livelihoods of Mongolian pastoralists. J Arid Environ 109:54–64. http://linkinghub.elsevier.com/retrieve/pii/S0140196314001256. Accessed 5 Oct 2014

Amgalanbayar T (2011) Mongolia: national progress report on the implementation of the hyogo framework for action (2009–2011). http://www.preventionweb.net/files/18911_mng_NationalHFAprogress_2009-11.pdf

Chai J (2015) Under the doom produced by ming fan CCTV. 2015.2.15

Chou H (2007) Introduction to Mongolia's local government system. Bimonthly J Mongolian Tibetan Curr Situat 16(6):1–47. http://www.mtac.gov.tw/mtacbooke/upload/09611/0101/e1.pdf. Accessed 20 Feb 2015

CNDS (2008) Comprehensive National Development Strategy of Mongolia, Government of Mongolia, Ulaanbaatar

Dagvadorj D et al (2009) Mongolia: assessment report on climate change 2009

ERST_UNDP_NEMA (2010) Dzud national report 2009–2010

Farber DA (2011) The challenge of climate change adaptation: learning from national planning efforts in Britain, China, and the USA. J Environ Law 23(3):359–382. http://jel.oxfordjournals.org/cgi/doi/10.1093/jel/eqr015. Accessed 8 Dec 2014
Japan (1994) Japan's action report on climate change
Japan (1997) Japan's second National Communication under the United Nations framework convention on climate change. http://unfccc.int/resource/docs/natc/japnc2.pdf
Japan (2002) Japan's third National Communication under the United Nations framework convention on climate change
Japan (2006) Japan's fourth national communication
Japan (2010) Japan's fifth national communication
Japan (2013) Japan's sixth national communication
Mayer B (2015) Climate migration and the politics of causal attribution: a case study in Mongolia. Migr Dev 1–20. http://www.tandfonline.com/doi/full/10.1080/21632324.2015.1022971
MEGD (2013) Technology needs assessment, vol. 1—climate change adaptation in Mongolia
Mimura N et al (2014) Adaptation planning and implementation. In: Contribution of working group II to the fifth assessment report of the intergovernmental panel on climate change. [Field (ed) Climate change 2014: impacts, adaptation, and vulnerability. Part a: global and sectoral aspects]. Cambridge University Press, Cambridge and New York, pp 869–898
MINC (2001) First National Communication to UNFCCC. Mongolia, 2001. http://unfccc.int/resource/docs/natc/mongnc1.pdf
MNEM (2001) Mongolia's initial National Communication to the UNFCCC. http://unfccc.int/resource/docs/natc/mongnc1.pdf
MNET (2010) Mongolia second National Communication under UNFCCC
MOE (2010) National Climate Change Adaptation Master Plan (2011–2015), Ministry of Environment, Republic of Korea
MOEJ (2008) Wise adaptation to climate change
MOEJ (2009) Synthesis report on observations, projections, and impact assessments of climate change—climate change and its impacts in Japan. http://www.env.go.jp/en/earth/cc/report_impacts.pdf and http://www.env.go.jp/earth/ondanka/rep091009/full.pdf
MOEJ (2010a) Approaches to climate change adaptation. http://www.env.go.jp/en/earth/cc/adapt_guide/pdf/approaches_to_adaptation_en.pdf
MOEJ (2010b) Climate change adaptation: approaches for national and local governments. http://www.env.go.jp/en/earth/cc/adapt_guide/pdf/approaches_to_adaptation_leaflet_en.pdf
MOEJ (2012) Climate change and its impacts in Japan
MoEJ (2015) National plan for adaptation to the impacts of climate change. https://www.env.go.jp/en/earth/cc/nationalplan151127-2.pdf. Accessed July 2017
National Development and Reform Commission (2007) China's national climate change programme, (June), p 63
NDRC (2010) White paper on climate change actions in China
NDRC (2013) China's policies and actions for addressing climate change (2013)
NDRC (2014) National Programmes on actions to climate change
NDRC (2015) Enhanced actions on climate change: China's intended nationally determined contributions. http://www.blackwell-synergy.com/doi/abs/10.1111/j.1467-9388.1992.tb00046.x
Pelling M (2011) Adaptation to climate change: from resilience to transformation. Routledge, Abingdon
Subbiah A et al (2013) Country report: climate risk management in Mongolia

Chapter 6
Status of Climate Change Adaptation in the Pacific Region

Espen Ronneberg and Peniamina Dougalii Leavai

Abstract Climate change and disaster risks increase the vulnerability of Pacific Island people. Climate change impacts also cause progressive long-term degradation to the natural environment, to critical ecosystems (e.g. coral reefs). They collectively undermine the sustainable development of the Pacific region significantly. This chapter highlighted the readiness of country and region and gaps between current level of adaptation practices and efforts at national levels primarily realised by the lead agency, as assigned by respective countries, and the efforts that are projected to reach a hypothetical 'safe' level of adaptation in view of recent science (i.e. AR5) and assessed needs for adaptation. The analyses are primarily based on eight indicators presented in the framework chapter.

Keywords Climate change · Adaptation · Resilience · Mainstreaming · Integration · Pacific region

6.1 Introduction

The Pacific Island region is comprised of 14 Small Island Developing States (SIDS), i.e. the Cook Islands, Federated States of Micronesia, Fiji, Kiribati, Marshall Islands, Nauru, Niue, Palau, Papua New Guinea, Samoa, Solomon Islands, Tonga, Tuvalu, and Vanuatu. Four are listed as amongst the least developed countries (LDCs) in the world: Kiribati, Solomon Islands, Tuvalu, and Vanuatu (Fig. 6.1). In 2014, Samoa graduated from LDC status to a Middle-Income Economy country. American Samoa, Guam, Commonwealth of the Northern Mariana Islands are territories of the United States. French Polynesia, New Caledonia, Wallis, and Futuna are territories of France, and Tokelau is an associated territory of New Zealand. The Pacific Island countries (PICs) referred to in this chapter will focus on 14 independent Pacific Small Island Developing States.

E. Ronneberg (✉) · P. D. Leavai
Secretariat of the Pacific Regional Environment Programme, Apia, Samoa
e-mail: espenr@sprep.org; peniaminal@sprep.org

M. Alam et al. (eds.), *Status of Climate Change Adaptation in Asia and the Pacific*, Springer Climate, https://doi.org/10.1007/978-3-319-99347-8_6

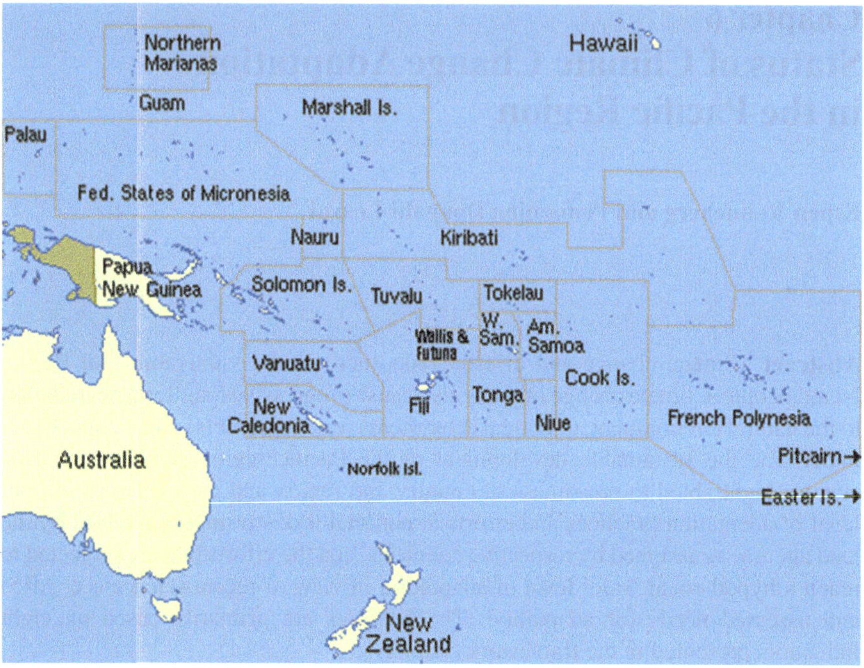

Fig. 6.1 Location of 14 small islands developing states in the Pacific

The features and size of Pacific SIDS are diverse and vary considerably although most of the countries are very small. The 14 SIDS have a land area of only 550,073 km^2 but are spread out in the world's largest ocean. Nauru and Tuvalu, for example, comprise a land area of less than 27 km^2. The large expanses of water and small areas of land create numerous challenges including difficulties of transport and communications both within and between countries and to the international community (PIFS 2014).

Many PICs have experienced the rapid growth in population in past decades along with increasing commercialisation of traditional and subsistence-based economies. This commercialisation process resulted rapid increases in rates of natural resource exploitation, especially land, forests, and living marine resources. On the other hand, many PICs are characterised by extremely small land area and limited land resources, such as soil and forest, making many terrestrial and near-shore resources quite vulnerable to overexploitation and pollution from poorly planned waste disposal system.

All PICs have a traditional dependence on marine resources for subsistence and as the region's ocean resources contain the highest marine biodiversity in the world. It represents almost the sole opportunity for substantial economic development for many of the Small Island Nations (SPREP 2009).

Many of the Pacific Islands, especially atolls, have very low topography. Some reach only a few metres above sea level at their highest point. In recent times, PICs often experience enormously damaging extreme weather events, suffer severely from global economic pressures and fluctuations, and have a heavy reliance on the productivity of one or two economic sectors. The 2008 and 2014 king tides, and 2015 tropical typhoons Maysak and Chan Hom caused major damages to people's homes and properties including government wharfs and ships in the north Pacific Island countries of the Marshall Islands and the Federated States of Micronesia. The Solomon Islands registered a first tropical cyclone to occur within the month of July 2015, the middle of dry season. These extreme events are exaggerated with the accelerated rise in sea level now experienced by communities who make these islands their home.

The small size of islands results in the interactions between components of natural systems and the sectors of economic activity being rapid and strong, which demands that sustainable development be holistic and balanced. This feature of smallness, highly integrated environmental and economic systems.

The geographic nature of the Pacific region causes isolation from the main centres of trade such as Asia, Europe, and the United States, creating unique difficulties in integrating into the global economy. Transportation costs reduce competitiveness, and uncertain air and shipping linkages are significant obstacles to efficient export manufacturing. However, the benefits of trade liberalisation and globalisation can only be realised in the Pacific region if the specific limitations and vulnerabilities of SIDS are addressed at all levels.

The countries in the region have been reporting serious socio-economic, environmental, physical, and cultural consequences of climate change. Numerous studies suggest that climate variability and change are likely to accentuate the spatial and temporal variations, including variability, which result from the El Niño Southern Oscillation (ENSO) events. The risks and impacts of climate change and variability for Small Island Developing States (SIDS) including the Pacific can be summarised as follows (IPCC 2015).

- *Sea level rise*: with more than 70% of the populations living in the coastal zone, at less than a few metres above sea level, Pacific Island countries are very vulnerable to the predicted levels of sea level rise.
- *Extreme weather events*: the higher category cyclones appear to be becoming more frequent, while the number of cyclones remains stable.
- *Changes in rainfall patterns*: projections indicate more droughts and more intense rainfalls in parts of the region, with implications on agriculture and human health.
- *Coral reefs*: increased sea temperatures and ocean acidification will have severe impacts on the health and viability of coral reefs—a key resource for tourism as well as for near-shore fisheries.
- *Threat to fisheries*: a shift in tuna migratory patterns has been documented due to El Nino conditions. This is a proxy for future climate change with impacts on individual country's economic returns from fisheries, food security as well as the longer term viability of Pacific fisheries.

- *Threats to human settlements and infrastructure*: about 80% human settlement and infrastructure are concentrated within the coastal zone. Given the limited ability of space to relocate, the costs to protect or rebuild will be immense for the Pacific. Insurance as risk sharing mechanism is often lacking, a parametric insurance scheme is being trialled in the region by SPC. It remains clear that insurance premiums will require subsidies, most likely from development partners and other climate finance mechanisms.
- *Consequences to human health*: changes in rainfall patterns and higher temperatures will affect prevalence of vector-borne diseases and their range, and will affect human populations through heat stress.
- *Enormous impacts on economy*: overall impacts of climate change on economy of the small islands countries in the Pacific region can be of a magnitude equivalent to their GDP or more.

6.2 Regional Adaptation Objectives

Climate change issues have attracted intense attention in the four atoll countries, i.e. Kiribati, Marshall Islands, Tuvalu, and the Territory of Tokelau due to their sheer vulnerability to sea level rise and existential threats from climate change impacts. Governmental authorities and scientific researchers have made a concerted effort in each nation to assess the negative impacts of CC. For example, Tokelau, with its miniscule emissions, decided to introduce 100% photovoltaic electricity supply in 2015–2016, and is also taking part in a pilot project on ocean acidification. The national visions and action plans on both mitigation of, and adaptation to, CC are summarised in Table 6.1.

The PICs established initial adaptation needs in their National Communications and have expanded on their ability to build and improve their adaptation programmes in the countries over the last two decades. From the mid to late 1990s, the PICs had focused their objectives to adapt to climate change largely on enabling and capacity building activities, carrying out baseline vulnerability and adaptation assessments, simple climate modelling and establishing ad hoc institutions (SPREP 2002). This changed quickly to implementing immediate and urgent adaptation activities from as early as 2002 when national communication reports to the United Nations Framework Convention on Climate Change (UNFCCC) recommended immediate actions for adaptation, especially for LDCs in the Pacific. A key policy document was developed for the region in the 2006–2015 Pacific Islands Framework for Action on Climate Change (PIFACC). The PIFACC is the regional climate change policy endorsed by the Pacific Leaders in 2005. The Framework's vision is 'Pacific island people, their livelihoods and the environment are resilient to the risks and impacts of climate change'. All PICs adhere to the following adaptation objectives as agreed under PIFACC:

Table 6.1 Adaptation objectives of various Pacific Island countries as stated in nationally available documents

Name of country	CCA objective(s), as stated in regional and national documents	Timeline/ interim timeline
Pacific Island region	*Vision:* Pacific Island people, their livelihoods, and the environment resilient to the risks and impacts of climate change *Goal:* To ensure Pacific Island people build their capacity to be resilient to the risks and impacts of climate change with the key objective to deliver on the expected outcomes (SPREP 2011)	2006–2015
Cook Islands	To consider implications of increasing risks associated with extreme weather and climate events. This is through taking an integrated approach to adaptation and disaster risk reduction, at institutional, policy, and practical levels *Development sectors prioritised:* coastal zones and infrastructure, agriculture, marine resources and fisheries, biodiversity, water resources, human health, and well-being (GOCI 2011)	2011–2015
Federated States of Micronesia	To (a) account for projected climatic changes in development activities and projects; (b) use ecosystem-based approaches where applicable; (c) encourage and strengthen the application of traditional knowledge on conservation practices and other relevant areas; and (d) develop and implement appropriate strategies to improve food production and other relevant sectors (GOFSM 2009)	No specific timeline is mentioned Policy endorsed in 2009
Fiji	To reduce the vulnerability and enhance the resilience of Fiji's communities to the impacts of climate change and disasters (GORF 2012)	No specific timeline is mentioned Policy endorsed in 2012
Kiribati	To communicate in a simplified way the identified immediate and urgent adaptation needs of Kiribati, which is also relevant to the national communication obligation required by the UNFCCC. These adaptation needs are identified through a participatory, consultative, and multidisciplinary planning process (GOK 2007)	No specific timeline is mentioned NAPA endorsed in 2007
Marshall Islands	To develop and have in place a contingency/adaptation plan to counter emerging threats resulting from the adverse effects of climate change, including a National Disaster Plan (RMI 2014)	2015–2017
Nauru	To identify effective solutions that support adaptation and reduce risks in the areas of water security, energy security, food security, the natural environment, health, and land resources (GON 2015)	No specific timeline is mentioned Policy Framework endorsed in 2014

(continued)

Table 6.1 (continued)

Name of country	CCA objective(s), as stated in regional and national documents	Timeline/ interim timeline
Niue	To develop effective adaptation responses and enhance adaptive capacity in order to protect livelihoods, natural resources and assets, and vulnerable areas to the impacts of climate change to all sectors (GON 2009)	No specific timeline is mentioned Policy endorsed in 2009
Palau	To achieve a resilient, sustainable, and food secure Palau (GOP 2015)	No specific timeline is mentioned Strategy endorsed 2015
Papua New Guinea	To achieve a climate-resilient and carbon neutral pathway (PNG 2014)	Long-term vision 2050 Policy endorsed in 2014
Samoa	To (a) develop and implement immediate and urgent project-based activities to adapt to climate change and climate variability; (b) protect life and livelihoods of the people, infrastructure, and environment; (c) incorporate adaptation measures and goals into national and sectoral policies, and development goals; and d) increase awareness of climate change impacts and adaptation activities in communities, civil society, and government (GOS 2005)	No specific timeline is mentioned NAPA endorsed in 2005
Solomon Islands	To provide a guiding framework to (a) integrate climate considerations and support the implementation and achievement of Solomon Islands National Development Strategy and other regional and international policies and frameworks and (b) guide the government and its partners' efforts in ensuring that the people, natural environment, and economy of the country are resilient and able to adapt to the predicted impacts of climate change (GOSI 2009)	2012–2017 Policy endorsed in 2012
Tokelau	To build Tokelau's resilience to a changing climate (TNCCS draft). It has also introduced 100% photovoltaic electricity for all three atolls. 2015–2016	2010–2015
Tonga	To promote and ensure safe, healthy, secure, and resilient communities to climate change impacts and disaster risks (KoT 2010)	2010–2015
Tuvalu	To protect Tuvalu's status as a nation and its cultural identity and to build its capacity to ensure a safe, resilient, and prosperous future (GOT 2012)	2012–2021

(continued)

Table 6.1 (continued)

Name of country	CCA objective(s), as stated in regional and national documents	Timeline/ interim timeline
Vanuatu	To develop a country-wide programme of immediate and urgent project-based adaptation activities in priority sectors, in order to address the current and anticipated adverse effects of climate change, including extreme events (GOV 2005)	No specific timeline is mentioned Policy endorsed in 2005

(a) Adaptation measures to the adverse effects of climate change developed and implemented at all levels.
(b) Identification of vulnerable priority areas/sectors and appropriate adaptation measures using available and appropriate information recognising that such information may be incomplete.
(c) Adaptation measures in vulnerable priority areas supported by existing data sets and traditional knowledge, or new data developed in some instances as necessary.
(d) Appropriate adaptation measures integrated into national/sectoral sustainable development strategies or their equivalent and linked to the budgeting process.

This could involve several activities such as (a) clearly identifying national adaptation priorities, (b) engaging with stakeholders in priority sectors to discuss impacts and appropriate adaptation responses, (c) developing national adaptation plans based on the priorities identified, with appropriate support from regional organisations and development partners, (d) designing national adaptation programmes, with appropriate support from regional organisations and development partners to implement these plans that address underlying vulnerabilities and support resilience building, and (e) committing national budgets for adaptation programmes as appropriate.

The early experiences from individual country projects revealed that implementation of adaptation projects faced financial and resource challenges which affecting sustainability and impacting on adaptation efforts in the countries. With ample capacity building activities from the late 1990s into the early 2000, the objectives have pushed for immediate implementation on the ground and with isolated island communities. The following Table 6.1 provides a summary of the stated objectives of adaptation in the PICs.

The Cook Islands Joint National Action Plan (JNAP) 2011–2015 vision summarises it rather succinctly in three keywords: safety, resilience, and sustainable development futures. The objective of the largest adaptation project in the region, the Pacific Adaptation to Climate Change (PACC) is increasing resilience and reducing vulnerability of the PICs from impacts of climate change in three priority development sectors,[1] namely coastal management, water security, and food security

[1] PACC project, www.sprep.org/pacc.

and food production. Eight Pacific Island countries that include Fiji, FSM, Marshall Islands, Niue, Samoa, Solomon Islands, Tuvalu, and Vanuatu developed policies on climate change that clearly outlined national adaptation priorities, providing leeway for all stakeholders to engage with each other in priority sectors to discuss impacts and appropriate adaptation responses.

Tonga, Cook Islands, Marshall Islands, and Niue have developed their JNAPs that integrate disaster risk reduction and climate change adaptation. The two former countries have completed implementation of the first period of their JNAPs. The Cook Island JNAP's (2011–2015) vision is 'a safe resilience and sustainable Cook Islands'. The adaptation objectives of its fourth strategic area that focuses on 'risk reduction and climate change adaptation' are (a) to strengthen infrastructure and safeguard essential services considering current and anticipated climate change; and (b) to strengthen economic development and livelihood systems in key sectors, increasing resilience to disasters and climate change.[2]

The Pacific LDCs have and continue to focus on the generic objectives set out by the National Adaptation Programme of Actions (NAPAs) since 2002, closely following the annotated guidelines of the NAPAs developed by the LDC Expert Group under the United Nations Framework Convention on Climate Change (UNFCCC COP Decision 28/CP.7). The objective of the NAPAs was to develop adaptation needs that are urgent and immediate. Vanuatu, Tuvalu, Solomon Islands, Kiribati, and Samoa (which graduated from LDC status in 2014) completed the development of their NAPAs between 2005 and 2009.[3]

While the LDCs clearly outlined their urgent and immediate priorities under the NAPAs, the PICs have moved to an integrated approach that ensured grass roots community involvement, funding security, and with a movement to quick implementation. Samoa, for example, reviewed its nine NAPA priorities that resulted in five integrated project programmes. Because of this integrated adaptation objective approach, the PACC funded Samoa's NAPA priority on coastal zone management. Fiji's adaptation objective under its policy is explicit with merging disaster risk reduction and adaptation and states that it is 'to reduce the vulnerability and enhance the resilience of Fiji's communities to the impacts of climate change and disasters'.[4] Fiji has since demonstrated climate change adaptation activities under the agriculture sector through the PACC project.

To achieve adaptation visions and objectives stated above, the Pacific Island countries have identified several measures which are centred around six key thematic elements. Table 6.2 summarises these elements which include integration, policy development, and policy support as one common mainstreaming goal. The other key elements are communication and advocacy; risk reduction; sector capacity development; community focused; and protection and conservation.

---

[2] Cook Islands JNAP 2011–2015.

[3] Samoa Dec 2005, Kiribati Jan 2007, Tuvalu May 2007, Vanuatu Dec 2007, Solomon Islands December 2008 (source: http://unfccc.int/adaptation/workstreams/national_adaptation_programmes_of_action/items/4585.php).

[4] Government of the Republic of Fiji, Republic of Fiji National Climate Change Policy, 2012.

Table 6.2 Key elements of Pacific Island country adaptation objectives

Key elements of adaptation objectives and visions	Countries specific to key elements	Key documents
(a) Integration, Policy Development, and Support: to provide, guide, and promote policy and decision- making through an integrated approach of climate change adaptation and disaster risk reduction	Pacific region, Tonga, Republic of Marshall Islands, Cook Islands, Niue, Tuvalu, Nauru, FSM, Samoa, and Vanuatu	Pacific regional strategy (in draft), SRDP, country JNAPs, climate change policies, NAPAs
(b) Communication, Advocacy: to communicate immediate and urgent needs and raise awareness	Vanuatu, Samoa, Kiribati, Solomon Islands, Tuvalu	NAPAs
(c) Community focused: to emphasise the process of identification of needs (being community, CSO, NGO, private sector, etc. consultations)	Vanuatu, Samoa, Kiribati, Solomon Islands, Tuvalu, Tonga	NAPAs, climate change policies, JNAPs
(d) Risk reduction: to respond effectively to emerging risks and threats with clear linkages	Cook Islands, FSM, Kiribati, Niue, Nauru, Tuvalu	NAPAs, JNAPs
(e) Sector capacity development: to develop adaptation capacity of sectors	Cook Islands, FSM, Palau, Nauru	National and state sector climate change adaptation policies, JNAPs
(f) Protection and Conservation: to protect life, livelihoods, infrastructure, and environment, and promote ecosystem-based approaches	Samoa, Kiribati, Solomon Islands, FSM	Legislation, Regulation including by-laws, social and environment safeguards

It is important to note that while these key elements apply generally to all SIDS in the Pacific given their geographical island settings, similar social, economic, and environmental circumstances (e.g. being LDCs), some of the countries have focused on specific factors largely because of their current state of readiness. For example, Samoa, Kiribati, Solomon Islands, and FSM place special emphasis on protection and conservation of its environment, people, and infrastructure and livelihood sources by promoting ridge to reef and ecosystem-based approaches.

It is clear from Tables 6.1 and 6.2 that the PICs have generally identified village and island communities as a key vulnerable sector that requires adaptation and is therefore the basis of many of the adaptation objectives of the countries. The women, children, and men of the communities are beneficiaries of these adaptation actions and are therefore the target groups of adaptation projects and programmes in the countries. The Cook Islands, Fiji, Kiribati, Palau, Samoa, Solomon Islands, Tonga, Tuvalu, Tokelau, and Vanuatu have all but focused their adaptation objectives around addressing community risks, vulnerabilities, and their ability to adapt.

The key priority development sectors that support and are linked in supporting the development of community capacities to adapt include water, health, livelihoods and well-being, coastal, agriculture, marine, fisheries, biodiversity, environment conservation, traditional knowledge, food production, infrastructure, and land resources. Three of these priority development sectors were demonstrated and implemented under the PAC project from 2009 until June 2015. These sectors are coastal zone management, water resource management, and food security and food production.

While countries such as the Cook Islands, Tuvalu, Vanuatu, and Solomon Islands are slowly progressing to articulate public expenditures to reflect climate finances, the capacity of governments to meet basic fiduciary standards required to provide a robust account of climate finance is an area that countries will take years to develop.[5] Samoa is an exception in that as it is fast practising commitment of national budgets for adaptation programmes as appropriate. It has set up its Climate Investment Unit under the Ministry of Finance as part of its institutional strengthening approach to climate finance management response to climate change programmes in the country.[6]

The national climate change policy of Solomon Islands is unique in comparison to other national climate change policies, in that it stresses increasing the resilience of its economy to be able to adapt to the predicted impacts of climate change.[7] The government promotes its commitment to invest in nationally appropriate mitigation actions (NAMAs) to reduce its own GHG emissions '…that brings benefits to the country's economy, environment and improves livelihoods of its people[8]'.

It is important to note that although the SIDS in the Pacific contribute only 0.002% of global greenhouse gas emissions, the Pacific Island countries nonetheless have, for the past two decades applied a dual approach to addressing climate change. The implementation of adaptation and mitigation activities has shaped the climate policy responses of many of the SIDS in the region. For example, Samoa, Tuvalu, Kiribati, Tonga, and Fiji promoted the use of renewable energy technology to reduce the reliance on imported diesel, saving electricity and cushioning the imported fossil fuel costs on the economy. Using solar energy to drive the generators that pump up water in the rural and outer island communities is popular in atoll and flat limestone islands such as Tonga, Nauru, and the Marshall Islands. This saves fossil fuels and has also allowed for stricter controls on pumping rates to be implemented as an adaptation measure.

[5] Personal communications with country representatives from ministries of finance, climate change offices, foreign affairs and trade at the Green Climate Fund workshop, Nadi, Fiji 27–29 July 2015.

[6] Ministry of Finance, Government of Samoa, 2015.

[7] Solomon Islands National Climate Change Policy, 2012–2017.

[8] Solomon Islands National Climate Change Policy, 2012–2017, (8.4.1.4 Policy Directive and Strategies, p 23).

6.3 Adaptation Trend

6.3.1 Policy, Strategy, Legal Provision, and Governance

Most PICs integrate climate change adaptation and disaster risk management into their national policies, development strategies and legal and governance frameworks. These are evident within social and economic development planning processes and practices that have already been undertaken ranging from the community level to regional level. In 2010, all PICs launched the Pacific Framework for Action on Climate Change (PIFACC) to guide its work on climate change, which ended in 2015. The Pacific is now the first region in the world to decide on a single integrated regional strategy to address climate change and disaster risk management. The strategy is to succeed the PIFACC 2010–2015 and the Pacific Disaster Risk Reduction and Disaster Management Framework for Action (RFA) that also ended in 2015. The Strategy for Resilient Development in the pacific is an integrated approach to Climate Change and Disaster Risk Management (SRDP) 2017–2030. It aims to provide targeted strategic guidance to different stakeholder groups on how to enhance resilience, in ways that contribute to the more fundamental goal of sustainable development. The strategic guidance and policy direction formulating the adaptation trend in the Pacific is therefore one of integration of climate change and disaster risk management.

Eleven PICs have already developed Joint National Action Plans (JNAPs) that integrate climate change and disaster risk management at the national level. The countries are Cook Islands, Federated States of Micronesia, Marshall Islands, Nauru, New Caledonia, Niue, Tonga, Tokelau, Tuvalu, Tonga, and Vanuatu. The plans are both policy and strategic planning documents that guide countries in addressing both climate change and disaster risk management at the national and sector levels.

Tonga—A Leader in Joint National Action Plans

The Kingdom of Tonga was the first country to formally combine work on climate adaptation and disaster risk reduction in the Pacific countries. In July 2009, the cabinet gave approval for the Ministry of Environment and Climate Change (MECC) to work with other ministries, NGOs, statutory boards, and development partners to develop the JNAP on Climate Change Adaptation and Disaster Risk Management 2010–2015. The preparation of the plan was deliberately scheduled at the same time as a country-wide consultation to develop Tonga's second national communication to the UNFCCC, reducing 'workshop fatigue' amongst NGOs and communities who participated in both meetings.

Table 6.3 Summary of institutional arrangement in Pacific Island countries

Country	Political	Operational lead	Climate finance	Adaptation/ mitigation programmes
Cook Islands	Office of Prime Minister	Office of Prime Minister	Ministry of Finance	National Environment Services
Federated States of Micronesia	Department of Foreign Affairs	Office of Environment and Emergency Management (OEEM)	Department of Foreign Affairs, OEEM	OEEM
Fiji	Ministry of Foreign Affairs	Ministry of Environment	Ministry of Finance	Sector ministries
Kiribati	Ministry of Foreign Affairs	Ministry of Environment	Ministry of Finance	Sector departments
Marshall Islands	Department of Foreign Affairs	Office of Emergency Policy, Planning Communications (OEPPC)	Office of Emergency Policy, Planning Communications (OEPPC)	Office of Emergency Policy, Planning Communications (OEPPC)
Nauru	Department of Foreign Affairs	Department of Commerce, Industry and Environment	Ministry of Finance	Sector departments
Niue	Office of Prime Minister	Environment	Ministry of Finance	Environment
Papua New Guinea	Ministry of Foreign Affairs	Office of Climate Change and Development (OCCD)	Office of Climate Change and Development (OCCD)	Office of Climate Change and Development (OCCD)
Palau	Office of the President	Office of the President	Bureau of Budget and Planning	Sector departments
Samoa	Ministry of Foreign Affairs and Trade	Ministry of Natural Resources and Environment	Ministry of Finance	Adaptation—MNRE Mitigation—MOF
Solomon Islands	Ministry of Environment, Climate Change, Disaster Management and Meteorology (MECDM)	MECDM	MECDM	MECDM
Tokelau	Government of Tokelau, Tokelau-Apia Liaison Office (TALO)	TALO	TALO	TALO

(continued)

Table 6.3 (continued)

Country	Political	Operational lead	Climate finance	Adaptation/ mitigation programmes
Tonga	Ministry of Foreign Affairs	MEIDECC	Finance—MOFNNP NDA GCF—MEIDECC	MEIDECC
Tuvalu	Office of Prime Minister	Office of Prime Minister	Office of Prime Minister	Department of Environment
Vanuatu		National Advisory Board	National Advisory Board	Ministry of Environment and Meteorological Services

6.3.2 *Institutional Arrangement*

Pacific Island governments continue to make changes to in-house arrangements in facilitation and coordination of their climate change programmes and projects. Most climate change focal points of Pacific Island governments are positioned within their central government agencies. These include the Ministry of Finance (Samoa), Office of Prime Minister (Cook Islands), Office of Economic Planning, and Office of the President (Palau, Marshall Islands). These are mainly political focal points. The operational focal points tend to be housed with line ministries that include environment, climate change, and meteorology. Table 6.3 below provides detailed information about institutional arrangement in Pacific Island countries.

The Pacific Islands have been building its capacity to coordinate and develop climate change programmes since the Pacific Island Climate Change Assistance Program (PICCAP) (SPREP 2009).

Papua New Guinea's experience of establishing a coordinating body for climate responses has been far more problematic than in other Pacific states, reflecting a complex interplay of factors. These include the diversity and size of an island nation of seven million people, the greater focus on reducing carbon emissions from deforestation and degradation of forests (REDD+) rather than adaptation, and broader problems in the management of public finances and lack of outreach to rural areas, at a time of major political change.

6.3.3 *Integration and Mainstreaming*

As with other Pacific countries, Vanuatu has begun to investigate the integration of climate change and disaster risk management into one structure. According to an April 2012 government report, 'the current separation between climate change and disaster risk management has led to stakeholder confusion, duplication of efforts, excessive time and resource demands on a limited number of actors, weak governance arrangements and a less effective delivery of services to ni-Vanuatu people'. The process of integration involves a range of tasks:

(a) Combining a series of overlapping strategic policy documents, including Vanuatu's UNFCCC National Adaptation Plan of Action (NAPA), the National Climate Change Adaptation Strategy (NCCAS), and National Action Plan on Disaster Risk Reduction and Disaster Management 2006–2016.
(b) Creating unified governance structures and processes. Until now disaster work has been managed through the Disaster Risk Reduction and Disaster Management National Action Plan Task Force (NTF), housed under the Ministry of Internal Affairs. In contrast, the body that oversees climate change implementation, the National Advisory Committee on Climate Change (NACCC), is located under the Ministry of Infrastructure and Public Utilities, chaired through the Vanuatu Meteorological and Geohazards Department.
(c) Passing new legislation beyond the existing National Disaster Act of 2000 to fully enable climate change and disaster risk integration (in contrast to the NTF, NACCC currently operates without formal legislation).
(d) Establishing a system of monitoring, reporting, and verification on adaptation initiatives.

The reform process is supported by locally based consultants funded by the World Bank and jointly by the Secretariat of the Pacific Community (SPC) and the German Agency for International Development (GIZ), as well as technical advice from the Secretariat of the Pacific Regional Environment Programme (SPREP), SPC, and other regional networks.

In February 2012, the Vanuatu government held a workshop to discuss the shift to a new coordinating structure. The aim was to effectively merge the structures that currently manage work on disaster and climate responses into a single national advisory board (NAB). This board officially came into existence in May 2012, replacing the NACCC, and is made up of senior officials with clear tasks and mandates, detailed in legislation. A fully staffed Project Management Unit is being created as the NAB secretariat, based at the Vanuatu Meteorological and Geohazards Department. The government has explicitly requested public feedback on the proposed structures, as well as draft legislation to implement the changes.

Strengthening partnerships and coordination, Vanuatu civil society organisations have formed a Climate Change Network, and local and international NGOs are working together on disaster responses through the Vanuatu Humanitarian Team. This networking has allowed them to come together to present common responses to government, especially over proposals in the draft legislation that might restrict their work on the ground.

6.3.4 Implementation Capacity and Decision–Making Tools

Adaptation has been recognised in the region as a key priority by leaders and by communities around the Pacific since the 1992 Rio Summit. Some adaptation-related activities were carried out within the context of the Pacific Island Climate Change Assistance (PICCAP) project. While primarily focused as an enabling activity project some vulnerability and adaptation assessment training was provided to the PICs. This followed a model prepared by Waikato University, New Zealand and the University of the South Pacific, Fiji. Using simple simulations, the model allowed participants to make predictions on climate change impacts on vulnerable areas. Vulnerability assessments highlighted the following key sectors which have been affected by climate change and sea level rise: coastal zone and coral reefs; agriculture and food security; marine resources; water resources; and biodiversity. Some example of the effects included a decline in fruit crops production and low export sales due to drought and low rainfall in previous years, and loss of agricultural land due to intrusion of seawater through flooding, inundation, and coastal erosion especially in the atoll islands.

6.4 Current Gaps and Areas of Adjustment

The adaptation actions identified above are assessed using the assessment framework presented in Chap. 2 of this book for an in-depth analysis. Adequacy and effectiveness are also assessed as per this assessment framework. However, the assessment will indicate BAU pathway of adaptation objective(s) set aside by the regional countries. This analysis yields gaps in BAU/'maintenance' pathway for CCA.

Gaps are presented systematically using eight indicators and the sub-indicators described in the framework paper. The analyses are primarily based on qualitative judgement regarding adequacy, presence/absence of instruments and practices, quality of delivery, and effectiveness of measures and instruments. Quality of compliance is also treated for the maintenance gaps analysis. Table 6.4 below gives a broad indication of some of the shared challenges faced in the region, aligned with the assessment indicators. It should also be borne in mind that monitoring and evaluation work in the region is also faced with challenges in itself because of transportation and communications costs, as well as availability of human capacity.

Table 6.4 Shared challenges faced by Pacific Island countries towards delivering CCA

Assessment indicators	Shared challenges faced by Pacific Island countries towards delivering CCA
Policy, strategy, legal provisioning, and governance	Policy harmonisation involving other sector-specific development policies is taking place in some countries, but implementation in practice remains a challenge Project management of CCA involving project development to execution of projects are challenging due to human capacity constraints
Institutional arrangement	Proper institutional arrangements, including coordination mechanism, supervision, monitoring, and oversight roles in relation to CCA activities have been developed in many countries but are under-resourced
Integration and mainstreaming	Clear national vision towards integration and mainstreaming has been reaffirmed by many countries, but are again challenged by human resource constraints
Finance	Access to adequate climate change financing and frustration with fast start financing are prevalent issues and complaints. Human capacity constraints, information, and relevant skills are lacking
Implementation capacity	The current capacity in most cases has been below required level, which needs to be enhanced rather quickly to facilitate implementation of CCA activities
Technology	There exist challenges in terms of CCA technology development, with some projects being able to use low-tech approaches instead
Decision-making tool and institutionalisation	There are many relevant tools, but using these decision-making tools, is challenging in Pacific Island countries
Awareness and access to information	Public awareness is high, but needs to be enhanced with better quality information for communities to better participate in decision-making processes

6.5 Readiness for Enhanced Adaptation Actions

The readiness of country and region is highlighting approaches and gaps between current level of adaptation practices and efforts at national levels (primarily realised by the lead agency, as assigned by respective countries) and the efforts that are projected to reach a hypothetical 'safe' level of adaptation in view of recent science (i.e. AR5) and assessed needs for adaptation. The hypothesis is that much enhanced efforts than current levels of adaptation practices must be committed to achieve the future adaptation vision for the PICs.

The analyses are primarily based on eight indicators, as stated earlier. The same elements of judgments are applied against a much enhanced ('hypothetical' adaptation vision/pathway, projecting from the current level of adaptation efforts.

6.5.1 Pacific Capacity Building Needs

In an era of rapid change in international economy, characterised by the redirection of investment flows away from primary producing activities and the quick outflow of capital from troubled markets (e.g. the SE Asia financial crisis of the late 1990s), adjusting to external shocks and the associated management of vulnerability is a foremost challenge to the SIDS.

Responses to 'vulnerabilities' can be by building up resilience or by planning around hazards and risks. As the SIDS have limited ability to absorb natural and human-induced shocks, and limited resources—strategic assessment and forward planning of land use and development provides an effective and efficient means to address vulnerability. Planning systems provide the framework for pro-active measures as well as the vehicle for mainstreaming environment into economic development pursuits, the multi-use of data, provision of consistent guidance, and early participation of the community. They provide the means for instituting confidence in decision-making, certainty in processes and security in 'investment' all essential for sustainable development (overall vulnerability reduction management).

6.5.2 Leadership

The responsibility for leadership to reverse the current trend of growing vulnerability falls squarely in the domain of the government. A major challenge for the government is resolving the conflicting advice and requirements regarding the role of government and its corresponding size and structure. For example, on one side, governments of the SIDS are being urged to: implement decentralisation; reduce the size of the public sector; adopt the Agenda 21 principles of citizenry participation and consultation; and effectively participate in and fulfil reporting requirements under international agreements either linked to the rapid pace of globalisation or environmental degradation.

Additionally, international agreements on trade call on the SIDS to take certain policy decisions that are very likely to negatively impact revenues for the public sector and subsequent social investment, for example, the WTO requirement to reduce tariffs on imports to facilitate free flow of trade. In industrialised and industrialising countries, tariffs are a possible mechanism for protecting domestic industry and likely to negatively impact on free trade. In the SIDS, there are a very limited number of domestic industries increasingly foreign owned, and tariffs are the principal means of sheltering them form injuries of increased imports and for collecting revenue for the public sector.

Case evidence confirms that despite the growing recognition that the development challenge is different and more difficult because of the peculiarities of the SIDS, this is seldom considering when donors or international organisations enter into agreements. For example, the SIDS are being advised to privatise their relatively small utilities (water, electricity, and communications). As several Caribbean SIDS are finding out, this advice is resulting in escalating costs of utilities. In the case of electrical energy, increases in costs have resulted in significant loss of textile jobs to countries with cheaper energy.

6.5.3 Human Resources Development

High energy costs usually correlate with low wages. The question that arise is, given the unique nature of the electricity sector in the SIDS, was a model developed for the UK the most appropriate for the SIDS? Was there ever any thoughtful analysis done before this model was recommended to the SIDS, as the means of mobilising investments for electricity generation? The SIDS lack of capacity in the areas of water and energy limits their ability to conduct the in-depth analysis necessary in most of cases to determine whether what is being prescribed is likely to have the desired outcome. Clearly, the SIDS need additional professional will, a wider or higher range of skills than is now available in the SIDS because of a number of reasons ranging from level of remuneration, career prospects, and educational infrastructure.

Reducing the overall vulnerability of the SIDS will require governments to have access to requisite capacity to analyse external advice and conditionality. Given the limited capabilities of the private sector in most of the SIDS, the minimal flow of FDI which is strongly linked to the perceived vulnerability of the SIDS, the limited market infrastructure in place that does not yet allow effective and efficient operation of market forces, governments will have to play a greater not reduced role to lessen the growing vulnerability of the SIDS. This opinion is inconsistent with the prevailing ideas of some development partners and international development organisations as to the size, structure, and function of government. Reducing economic vulnerability will require the SIDS to reduce their dependence on imported fossil fuel, undertake economic diversification, and protect the biological resources.

6.5.4 Institutional Strengthening

As development assistance becomes less available, it increasingly comes with more conditionality that represents donor's perspectives on how beneficiary countries should pursue sustainable development. The SIDS will need the capacity to do the necessary analysis to determine relevance as well as the best methodology for implementation. The international community should give priority to addressing these aspects of need in order to help SIDS more effectively and efficiently reduce vulnerable and build resilience. Consider the case of food security. The vast majority of resources devoted to agriculture in the SIDS is directed at the production of commodity crops (sugar, bananas, copra, coffee), a legacy of the colonial history. Over the past decades, the unit value has continued to decline as once lucrative preferential agreements get dismantled by the coming into force of WTO agreements.

The dismantling of these historical agreements without implementing alternatives is resulting in increased vulnerability of the farmers and their households. The current strategy for food security advocated by technical assistance agencies like the World Bank is for the SIDS to depend on the international market. A strategy complimented by the WTO who wants the SIDS to lower tariffs on imported food. Does this advice reflect the vulnerable status of the SIDS economy or its farmers, and will the long-term impact lead to improved food security, sustainable natural resource use or is it likely based on present trends to result in further environmental degradation and ecosystems loss and increased overall vulnerability? The SIDS institutional capacity needs to be strengthened to address the interrelated nature of the use of resources, social outcomes, and environmental impacts.

Reducing environmental vulnerability in SIDS will require provision of the necessary social, educational, and economic support for the protection of natural resources on which all development is predicated in SIDS. In small societies, especially where the domestic sector is weak and undeveloped, the role of the Government must, of necessity, be more interventionist than it would be in large and more developed economies. A major challenge of the government as it seeks to reduce national vulnerability will be increasing the efficiency and effectiveness of the government. The government will have to do a better job of differentiating causes and symptoms, identifying linkages, and optimising the potential returns on limited investment resources. The framework for achieving much of this is detailed in the Barbados Plan of Action (BPOA) (for the BPOA as well as its successor frameworks MSI and SAMOA Pathway[9]. Unfortunately, as stated in the words of the UN Secretary General at the Special session on SIDS in 1999, limited capacity and resources has significantly hindered progress in implementing the BPOA by SIDS.

It is equally necessary to develop the competitiveness of the private productive enterprises for it is they who conduct the nation's trade. Competitiveness in the modern world demands a greater use of knowledge to inform decision-making, to provide services at international standards, and to utilise modern technologies in the production processes. It is also enhanced through niche market export strategy, flexible specialisation, enhanced entrepreneurship and, where appropriate, economic deregulation.

6.5.5 Local Level

The engagement of many SIDS governments, especially those in the Caribbean, in the range of functions that now exist, has its genesis in colonial history. The transition from colonial administration to self-government and finally to political

[9] See http://www.sids2014.org/.

independence heightened the consciousness of community leaders as to the social responsibility of government to provide basic services to a highly dependent population. Furthermore, the collapse of the plantation-based agricultural dominated economy left an extremely vulnerable and disorganised private sector dependent on protected markets, import trading and various forms of government intervention. Unlike other countries, the SIDS did not have the advantage of significant industrial progress before entering into a fully democratic and liberal system of governance, who today are among the most open global trade markets. It is not surprising therefore, that in many SIDS, governments have had to assume a role beyond that of regulator, to the point where their influence dominates every aspect of social and economic life, as the main employer in SIDS as well as being the main supplier of goods to outer islands and remote areas, and governments derive revenues from tariffs on trade.

Getting the policy right is proving to be very difficult. Policies being formulated seem to attract investment in one sector, tourism. Instead of economic diversification, the agreed-on approach for reducing vulnerability in SIDS, in actuality, most SIDS are fast-becoming tourism dominant economies with very little to supplement or take the shocks in case of negative downturns or during periods of economic difficulty. As SIDS's economic future becomes disproportionately more tourism focused, vulnerability increases. The growing vulnerability of SIDS is manifesting itself at the local level in turn of poor health and sanitation services, limited prospects for employment are proving difficult issues for national government to cope with. Innovative approaches call for diversification to the local level; however in the absence of capacity at this level, the anticipated result will not be realised. It was the view of SIDS therefore essential that the proposed SIDS Capacity 2015[10] programme should provide support for capacity building at the local level. However, this partnership is still dependent on resources. The private sector and NGOs/CBOs will all have to play larger role at the local level.

6.5.6 Capacity Building in SIDS for Vulnerability Reduction

The importance of capacity building for Small Economies has been emphasised in the Commonwealth Secretariat/World Bank Task Force report (Small States: Meeting Challenges in the Global Economy; report of the Commonwealth Secretariat World Bank Joint Taskforce on Small States, published by World Bank, 17 April 2000). Limited capacity has also been cited as one of the principal reasons responsible for the slow progress with the implementation of the BPOA. The Global Environmental Facility (GEF), in partnership with the UNDP, has devoted special attention to the capacity needs at various levels in SIDS for implementation of the global MEAs and to support capacity building for implementation of Agenda 21 generally, as well as the SIDS specific UN declarations. In addition to the

[10] See http://www.sids2014.org/.

implementation of Capacity 21 in some SIDS, the major capacity building activity is the recently completed national Communications reports under UNDP's Climate Change enabling activities.

Notwithstanding the general current trend towards leaner and meaner administrative structures and economic liberalisation, an increasing concern for the environment is bringing with it an increased requirement for regulatory activity. Equally, the forceful pressure towards consumer-oriented services is leading to requirements for more, not less government, with an emphasis on transparency rather than size. While this role for the government may run contrary to the prevailing liberal economic ideas on development, there is no market force available nor emerging in most of the SIDS to play an expanded role in reducing the vulnerability. The SIDS governments will need to convince donors of the uniqueness of their situation and the need for greater partnerships to develop capacity at all level of its civil society, private sector, and the government in order to meet the challenges of sustainable development starting with reducing economic, environmental, and social vulnerability in order to be more effective in the areas of water, sanitation, health and energy services, food security, and protection of biodiversity.

The need to rationalise public expenditure in situation of chronic budgetary deficits, to improve the quality of public goods and services, and to control increasing demand for such goods and services has brought into question the level of the government's capacity to continue to serve as the main and direct provider. In the long term, local groups will have to assume many of the function that the government by default has to play. However, for successful functioning at the local level community capacity has to be significantly strengthened.

6.6 Attaining Adaptation Vision

The lessons from temporary organisations such as the PACC project provide a significant learning curve for future regional and national adaptation programmes in the Pacific. The PACC has supported and itself set contemporary trends in adaptation within the countries. It has developed on-the-ground tools tailored to the Pacific, and ventured into real-time integration of policy and demonstration work on the ground and yielded successful results. It has shared successive failures as well, in a bid to promote a more effective and efficient way of implementing better adaptation projects that benefit village and island communities in the long run.

The following are some of the key PACC project lessons summarised with the view to share each of these country practices or procedures that are accepted or prescribed as being correct or most effective (i.e. best practices).

Cook Islands: Projects such as the climate proofing of coastal infrastructure should not be carried out as stand-alone efforts—they need to be part of a comprehensive coastal management planning within an integrated coastal management policy, and backed up by appropriate and effective legislation. Alongside the demonstration project, the PACC team has worked to develop a policy framework for coastal management on Mangaia Island. However, this also needs to be aligned with national coastal policy (which has not yet been formulated). Other requirements include appropriate governance structures that are well coordinated and that encourage wide participation of all stakeholder groups.

***Federated States of Micronesia*:** Good strategic planning and governance is essential to protect coasts and the FSM project demonstrated these well. For example, mainstreaming of climate change into national-level coastal policy and planning, and with communities to develop practical integrated coastal zone management plans and actions. The State of Kosrae found it crucial to support the development of the Kosrae State Climate Change Act, which was endorsed in 2011. The PACC lead the changes and amendments to Kosrae's Regulations for Development, which now require all development projects to consider the potential impacts of climate change. It looked at environmental safeguarding instruments such as the EIA Guidelines and incorporated climate risks into the procedures. Lastly, mainstreaming climate change risks into revised Kosrae Shoreline Management Plan that now provides a comprehensive strategy for building resilience of Kosrae's coastal communities and infrastructure into the future. These are examples of most effective practices for addressing climate change at the sector level plans and programmes in any country.

***Fiji*:** For any food security project efforts, farmers and farmer associations should be involved as much as possible. On-farm research, such as was carried out by the project, is the best approach for transferring any new methodology or practice. Further, the trial is relevant to that location, rather than the research station, which negates any genotype by environment effect. Working with the farmers, right from the start of such a project also provides the opportunity to address potential problems such as the non-marketability of any introduced crops and varieties.

***Republic of Marshall Islands*:** Technical assessments are essential for complex infrastructure projects such as the reticulation network in Majuro. The V&A process helped to identify issues with the network and possible adaptation options. From there, MWSC conducted various evaluations to assess the leakages of each storage tank and the loss of water through evaporation. The assessment of the reservoir from Fabtech later confirmed these results and evaluated the costs involved in the adaptation options. Finally, the use of a CBA provided important insight on how to maximise benefits within the given budget. These technical assessments allowed designing a tailored, relevant, and effective demonstration project.

***Nauru*:** Identify operating and maintenance needs during project design. Maintenance of water infrastructure is a major issue in Nauru. The overall state of the water network and storage is poor. At the household level, 50% of rainwater harvesting infrastructure is not properly maintained. It is important to fully consider these issues during project design, including running costs, funding, and operator capacity, and interest in maintaining the related infrastructure.

Niue**:** The PACC demonstration project in Niue exemplifies good coordination between the Government of Niue and donor partners. The strong commitment of the Government to an all-of-island approach and the flexibility of donor partners through the PACC, PACC+, and GCCA projects have contributed to the success of the project. Individually, none of the projects would have had sufficient funds to build the moulding factory and provide the entire island with rainwater harvesting systems. With the original PACC funding alone, the project would only have been able to cater for about 20% of the island's households, without the development of a permanent facility on island.

Palau**:** Sustainable farming systems—systems that maintain soil health, use water efficiently, respect and promote biodiversity, and produce good yields under current climate variability—are a vital base for building climate-resilient food production systems. As part of the drive to increase local food production, reduce reliance on imported foods, and address the non-communicable disease crisis, Palau promoted growing and eating local foods. The food processing component trained many stakeholders in local food processing and cooking, mostly with young people and 80% of participants being women. Sixty new recipes have also been developed, substituting locally grown ingredients instead of imported ingredients. These are 'no regrets' adaptation measures that include sustainable farming methods, climate-resilient crops, and training in food processing and preservation.

Papua New Guinea**:** In most isolated communities, low-tech, low-cost irrigation system which the farmers can easily set up and manage is a sustainable activity. This is because it utilises local materials that are available locally and can be sourced with incurring costs to be bought from distant urban areas. The community of Kivori in PNG is one of those isolated communities located about 5 h drive west of the capital. It has no electricity, no piped water and poor road infrastructures with no bridges. Thus, establishing supporting institution at the local level such as the Kivori Cooperative Society assists in ensuring the low-tech low-cost irrigation systems are functioning. They also help to support farmers to address challenges and to expand and reach new markets.

Samoa**:** Sea level rise, storms, high waves, and high tides are already damaging fragile and often densely populated coastal zones. With the added threat of climate change, measures are urgently needed to protect and manage the coasts. These must consider the natural systems, environmental change, and people's needs. A key lesson from the Samoa project is the importance of careful and thorough planning in the initial stages of adaptation projects, and the use of tools such as vulnerability and adaptation assessment, socio-economic assessment, and cost–benefit analysis to identify and evaluate all options to select the most appropriate one. Another lesson is that 'hard' adaptation measures need to be designed based on a detailed understanding of the surrounding coastal and watershed environment and flood conveyance routes. This should include coastal and flood risk modelling, and climate and sea level projections. A third lesson is the need for awareness raising amongst decision-makers, including communities, of climate change and its impacts, and especially the range of measures available to reduce vulnerability. When decision-makers understand the climate risk, and the options to reduce the risk, they will make informed and better decisions.

***Solomon Islands*:** In terms of mainstreaming climate change into national development plans, the project sought country-wide consultations that saw public and province consultations all the way down to community meetings. In soliciting views of what goes into and form the national climate change policy. The PACC was instrumental in drafting and then endorsement (in June 2012) of the national Climate Change Policy, which is now being implemented. One of those implementations targeted the hard to reach, isolated atoll islands of the Solomon Islands. Implementing food security to respond to impacts of climate change in these atolls, the PACC successfully developed demonstrations of the recommended permaculture systems. The systems are designed to be low-maintenance and high-yielding, and incorporate good management practices to restore soil health while producing good yields of diverse food crops. The project also set up a hybrid solar dryer in Honiara to demonstrate preservation of food by this method, which can then be shipped to atoll communities in times of need. Training in processing and preservation has been carried out, with a focus on women's groups. These practices have proven effective in the Solomon Islands and are encouraged for similar isolated islands of the Pacific.

***Tonga*:** The project team found that it is not easy to regulate common resources used by everyone, i.e. water in this case. People fear that their use will be restricted. The team addressed this by raising awareness of good water resources management, for fair use and sustainability, as provided by the national Water Policy. The lessons learned workshop was an extremely useful exercise for the project team. Bringing together diverse project stakeholders, it provided the opportunity to have open and honest discussions on what had worked and what could have been done better or differently. The lessons learned workshop is recommended as a best practice for similar projects.

***Tuvalu*:** Comprehensive planning can deliver great outcomes. It can be argued that the construction of a cistern is a relatively easy task compared to other PACC water projects; however, the PACC project Tuvalu has been very successful in identifying gaps in knowledge and necessary steps to implement its demonstration project. The project successfully identified one of the most vulnerable communities, selected a sustainable and efficient solution within the PACC budget, and used informal community leadership to implement and manage the adaptation measure.

***Vanuatu*:** Climate analysis is a key step in the early planning stages of climate-sensitive projects, alongside other essential assessments such as socio-economic assessment and cost-benefit analysis. Vanuatu demonstrated how to do this for coastal zone management projects, within a vulnerability and adaptation assessment framework engaging communities in owning the project outcomes and results through the training, hands-on participation, and use of the three-dimensional participatory mapping tool. The key objective is to help everyone with different levels of understanding to be on the same level of appreciation of their island or community environment. It therefore helps in decision-making. As a result, a 39 km stretch of road was selected for improvements and work is underway. The road improvement strategies are based on the results of the community mapping exercise, a vulnerability and adaptation assessment, and an environmental impact assessment. The design work has been completed, and the communities themselves have been trained to carry out the roadworks, using locally available materials and with guidance and training from the PACC project team.

6.7 Conclusion

Climate change and disaster risks increase the vulnerability of Pacific Island people, and significantly undermine the sustainable development of the Pacific region. The effects of climate change magnify disaster risks, for example, by increasing the frequency and/or intensity of extreme weather events. Climate change impacts also cause progressive long-term degradation to the natural environment, to critical ecosystems (e.g. coral reefs), and to social and economic systems, resulting in loss and damage to the system upon which Pacific Island communities depend on their subsistence and livelihoods. National and subnational governments and administrations, the private sector, civil society organisations, communities, and regional organisations and other development partners all have unique and key roles to play in addressing these challenges, individually and in partnership, to build a more resilient future for the Pacific region.

As a result, the Pacific region has developed 'The Strategy for Resilient Development in the Pacific: An Integrated Approach to Climate Change and Disaster Risk Management (SRDP)'. This provides high-level strategic guidance to different stakeholder groups on how to enhance resilience to climate change and disasters, in ways that contribute to and are embedded in sustainable development.

Part of the high-level strategic guidance provided through the SRDP is expressed in the form of non-exhaustive sets of 'priority actions', for consideration by the different stakeholder groups. These actions provide guidance only and are to be implemented as relevant to the individual priorities and needs of stakeholders. Some actions may be better implemented at the regional level and some would need to be further articulated at the national level to suit the specific context, priorities, and needs of each individual PICTs.

The SRDP advocates for the adoption of integrated approaches, whenever possible, for coping with and managing climate change and disaster risks, in order to make more efficient use of resources, to rationalise multiple sources of funding which address similar needs, and for more effective mainstreaming of risks into development planning and budgets.

Climate change and disaster risks are cross-cutting, and action must therefore take place at the sectoral level. Development sectors (such as health, education, water and sanitation, social assistance, energy, agriculture, fisheries, forestry, tourism, mining, culture, environment, and infrastructure) are recognised as having a particularly important role to play in owning and implementing resilient development measures. On top of that, the region has also developed cooperative arrangement for capacity supplementation and regional coordination, recognising that in some cases a national approach alone will not produce optimal results. This is part of the longest-term vision of the region to build resilience to climate change.

References

GOCI (2011) Nation communication. http://unfccc.int/national_reports/non-annex_i_natcom/items/10124.php

GOFSM (2009) National Climate Change Policy, Federated States of Micronesia. http://unfccc.int/national_reports/non-annex_i_natcom/items/10124.php

GOK (2007) National Adaptation Programme of Action (NAPA). Government of Kiribati. http://unfccc.int/adaptation/workstreams/national_adaptation_programmes_of_action/items/4583.php

GON (2009) National Climate Change Policy. Government of Niue. https://www.pacificclimatechange.net/node/55

GON (2015) Republic of Nauru Adapt. https://www.pacificclimatechange.net/node/54

GOP (2015) NSCRAA. Palau. https://www.pacificclimatechange.net/node/57

GORF (2012) National Climate Change Policy. Republic of Fiji. https://www.pacificclimatechange.net/node/9483

GOS (2005) National Adaptation Programme of Action (NAPA). Government of Samoa. http://unfccc.int/adaptation/workstreams/national_adaptation_programmes_of_action/items/4583.php

GOSI (2009) National Climate Change Policy. Government of Solomon Islands. https://www.pacificclimatechange.net/node/59

GOT (2012) National Climate Change Policy 2012. Government of Tuvalu. https://www.pacificclimatechange.net/node/62

GOV (2005) National Adaptation Programme of Action (NAPA) 2005. Government of Vanuatu. http://unfccc.int/adaptation/workstreams/national_adaptation_programmes_of_action/items/4583.php

IPCC (2015) www.ipcc.ch

KoT (2010). Joint National Action Plan on Climate Change Adaptation and Disaster Risk Management 2010–2015. Kingdom of Tonga. https://www.pacificclimatechange.net/node/61

PACC Experiences series 3–6. https://www.sprep.org/pacc/publications

PACC Nauru (2012), PACC Tonga (2014), Vital Harbor (2014), Vital Roads (2014). https://www.sprep.org/pacc/publications

PACC Technical report no. 1, 2, 3, 4, 5, 6, 7, 8, 9, 10, 11 and 12. https://www.sprep.org/pacc/publications

PACC Videos—Vital Health (2014), Vital Coast (2014), Vital Water (2012), Vital Foods (2012). https://www.sprep.org/pacc/publications/multimedia

Pacific-Australia Climate Change Science and Adaptation Planning Program (2014) http://www.bom.gov.au/climate/pacific/index-pacific.shtml

PIFS (2014) http://forumsec.org/resources/uploads/attachments/documents/Pacific%20experiences%20with%20modalities%20relevant%20for%20Climate%20Change%20Financing,%2020121.pdf

PNG (2014) National climate compatible development management policy (NCCDMP) is PNG's s blue print. https://www.pacificclimatechange.net/node/58

RMI (2014) National strategic plan 2015-17. Republic of Marshall Island, June 2014. https://www.pacificclimatechange.net/node/53

SPREP (2002) www.sprep.org

SPREP (2009) www.sprep.org

SPREP (2011) www.sprep.org

WB (2000) Commonwealth Secretariat/World Bank Task Force report - Small States: Meeting Challenges in the Global Economy; Report of the Commonwealth Secretariat World Bank Joint Taskforce on Small States, published by World Bank, 17 April 2000

Chapter 7
Status of Climate Change Adaptation in South Asia Region

Ahsan Uddin Ahmed, Arivudai Nambi Appadurai, and Sharmind Neelormi

Abstract Climate change is predicted to have major consequences for South Asia. The South Asian region represents the most diverse ecosystems, topographies and climate regimes in the world. Incidences of droughts, floods, heat waves and cyclones have grown both in terms of intensity and frequency, impacting the lives and livelihoods of the most impoverished and vulnerable people of South Asia. The regional countries and respective governments have all exhibited high level of political will and urgency to tackle climate change by means of adaptation, while committing to contribute towards achieving global mitigation goal if they receive adequate international supports in terms of finance, technology transfer and capacity building. Though several climate change adaptation (CCA) efforts are in place at the national and subnational levels in South Asia, they have so far been fragmented and incoherent lacking a perspective that integrates technological, institutional, financial, capacity, information and policy needs. Focusing on key countries of the region, this review captures the trends, strategies and critical barriers in advancing the CCA agenda. It argues for a regional vision and underscores the importance of cross-sectoral coordination, stakeholder integration, democratic decision-making, building synergies with local government institutions and enhanced capacities to tap financial resources from bilateral and multilateral funding agencies.

Keywords South Asia · Climate change · Adaptation · Readiness · Enhanced adaptation actions · Decision-making · Institutions · Vision · Nationally determined contributions

A. U. Ahmed (✉)
Centre for Global Change, Dhaka, Bangladesh

A. N. Appadurai
Climate Resilience Practice, World Resources Institute, Bengaluru, Karnataka, India

S. Neelormi
Economics Department, Jahangirnagar University, Dhaka, Bangladesh

M. Alam et al. (eds.), *Status of Climate Change Adaptation in Asia and the Pacific*, Springer Climate, https://doi.org/10.1007/978-3-319-99347-8_7

7.1 Introduction

South Asia[1] comprises Bangladesh, Bhutan, India, Maldives, Nepal, Pakistan, and Sri Lanka. The region represents the most diverse ecotones, topographies and climate regimes in the world (Muhammed 2003). It houses the largest delta on earth (i.e. the Bengal delta), the vast coastal plains and islands, peninsula, deserts, the tallest mountain peaks around the globe including the Himalayas, the largest mangrove forest, and a vast floodplain of a few of the most known rivers of the world such as the Indus, the Ganges, the Brahmaputra and the Meghna (de Sousa et al. 2009; Ahmad et al. 1994). The latter three collectively produce a huge outflow to the sea, only the second largest after the Amazon outflow. The Southwest Monsoon and the winter cold-stream across the Himalayan high mountains influence towards characterizing climate system in the region (Mirza and Ahmed 2003). While glacial melt from the Himalayas gives majority of the flow in the Indus River and other rivers of the Western Himalayan Region, the Monsoon runoff characterizes flow distribution in the Eastern Himalayan rivers. Since Monsoon is highly active only between June and first half of October, the rivers of the Eastern Himalayan Region (HER) exhibit acute distribution of water—'too much' of it during Monsoon creating frequent floods and 'too little' during the dry season reducing flow significantly, resulting in inward push of saline front from the ocean (Ahmed 2005; Ahmad et al. 1994).

The Monsoon front travels from east to west direction across the region, while the front recedes from west to east direction. This leaves little rainfall in the Western Himalayan rivers. Moreover, a large tract of land in western India and Balochistan Province of Pakistan falls in deserts (Muhammed 2003). Due to high evapotranspiration, the entire Western Himalayan Region (WHR) faces frequently occurring droughts (Mitra 2003). The food production system in Pakistan is completely dependent on canal-based irrigation system, which in turn is largely dependent on allocation of water in the Indus as per the 1962 Treaty on Sharing of Indus River Water between India and Pakistan (Ahmad et al. 2003). In the tails of the canal irrigation system, farmers often do not receive adequate amounts of water for irrigation. In the Southern Pakistan, water is also drawn from groundwater aquifers often beyond sustainable levels, which is why there has been a sustained salinization process in the groundwater aquifer system. Both drought and salinization of land are key climate driven hazards that are being faced by the people of the Western Himalayan Region (Mitra 2003; Ahmad et al. 2003). However, in recent decades, there have been repeated instances of occurrence of high intensity floods in parts of Pakistan (Wang et al. 2011; Tariq and van de Giesen 2012) and drainage congestion in congested urban areas in India due to excessive rainfall episodes (Ranger et al. 2011; De et al., 2013; Goswami et al. 2006; Pattanaik and Rajeevan 2010).

[1] For the present study, only seven countries out of a possible eight have been considered to delineate the South Asia region. Due to paucity of data as per the analytical requirement, Afghanistan had to be left out of the analysis. Afghanistan is very much part of South Asia and also a member of South Asian Association for Regional Cooperation (SAARC).

Seasonal flooding is a regular phenomenon in the Eastern Himalayan Region (Ahmed 2005; Mirza and Ahmed 2003). Terai region of otherwise mountainous Nepal, Indian states of Bihar, West Bengal, Meghalaya and Assam, and almost entire Bangladesh are subject to annual rise in water level in rivers, often leading to flooding (NCVST 2009; Dutta et al. 2013; Alam et al. 1998). However, in recent years flood intensity has increased, while the frequency of occurrence of high intensity floods has also increased significantly (Mirza and Ahmed 2010). In addition to flooding, the coastal plains of the sub-region have been subject to high intensity cyclonic storm surge, which often wreck havoc on human settlements and also on natural coastal ecosystems (Agrawala et al. 2003; Ahmed 2005; Nishat and Mukherjee 2013). In recent years, there have been a rise in the occurrence of 'low' and 'depression' along the Bay of Bengal because of a steady rise in Sea Surface Temperature (SST), which has been threatening lives of sea going fishermen and also severely affecting their livelihoods (Ahmed and Neelormi 2007). There have been reports of unusual high tides even during winter months, which are adversely affecting coastal agriculture and human habitats (MOEF 2012a, b). Moreover, due to lack of appreciable rainfall and runoff during non-Monsoon months, the Eastern Himalayan Region is also facing phonological moisture stress—the latter being demanding irrigation for the winter and summer crops (Mirza and Ahmed 2010).

The shift of temperature regime upwards in the mountains (in Pakistan, India, Nepal and Bhutan) has been affecting orchards and other agricultural practices (Regmi and Bhandari 2012). The natural streams are found to be drying up, or at least decreasing in volume which in turn has been adversely affecting potable water and agricultural practices in the mountains (Jianchu et al. 2007). Excessive rainfall episodes have already been threatening infrastructures and causing landslides (NCVST 2009). Rapid glacial melting has been creating Glacier Lake Outburst Floods (GLOFs) (Shrestha and Aryal 2011; Bajracharya et al. 2007), which might significantly decrease investment potential on infrastructure in mountainous areas (NCVST 2009). The islands (such as atolls in the Maldives and Sri Lanka) are facing coral bleaching, partial to complete inundation due to sea level rise and increased cyclonic vulnerability under increased SST (MOE 2011a, b; ROM 2001). The existence of the atolls in the Maldives is under significant threat as a consequence of global warming (ROM 2001, 2007).

In both WHR and HER, crop agriculture is the primary occupation of the inhabitants (Islam et al. 2011). The flood plains of the major rivers offer fertile lands which is suitable for staple grainy crops such as paddy and wheat (Muhammed 2003). Due to green revolution (extension of high yielding varieties, irrigation, investment in inputs such as fertilizers and pesticides), productivity of crops has increased significantly in recent decades. However, a large population is still trapped in acute poverty in the region (Lal et al. 2011; Kumar and Parikh 2001; Kelkar and Bhadwal 2007). They hardly can eke out a living from traditional agricultural practices (including fisheries, forestry and animal husbandry). As a consequence, the region houses the largest density of poverty ridden population in the world (Sivakumar and Stefanski 2011) and about 456 million people who have been facing malnourishment (FAO 2009). The fifth Assessment Report of IPCC indicates a rise in extreme weather events and subsequent devastations by the climate induced events as described above (IPCC 2014), which will most likely affect the poor and malnourished population of South Asia under climate change.

The above realities warrant immediate and profound adaptations to be initiated throughout South Asia. Regional countries have placed emphasis on sustaining agricultural growth and reducing vulnerability to water resources and water-related hazards (MOE 2011a, b; MOEF 2012a, b; GOIRP 2003; RGOB undated; ROM 2001; GON 2011a, b). Several countries have emphasized on planned approaches to address adaptation, while combining community-based approaches with institution led systematic adaptation (GOI 2009; GOP 2012; GOP/UNEP 1998; GED 2014; MOE 2011a, b). In the light of very high adaptation needs for the region, it is imperative to analyse the status of adaptation in South Asian countries, especially keeping in view the assessment framework provided in the previous chapter. As part of the assessment methodology, efforts are made to carry out the analysis based on the methodology discussed by Alam and Huq (2017), as placed in the Chap. 2 of this volume.

7.2 Adaptation Objectives of the South Asian Nations

South Asia as a region has not yet set any commonly applicable regional adaptation goal, even though each of the countries in the region has separately voiced their concerns related to vulnerabilities to climate change (RGOB undated; NATCOM 2004; ROM 2007; GON 2010, 2011a, b; MOE 2011a, b; MOEF 2009). Since South Asia has the largest density of food insecure people, food security has been the common key concern in the national statements on climate change (MOE 2011a, b; MOEF 2009, 2012a, b; GOIRP 2003; RGOB undated; ROM 2001; GON 2011a, b). Food insecurity and inadequate access to food by the poor and the marginal population has been given high attention in the official documents and strategies of Bangladesh, Bhutan, India, Nepal and Pakistan (ROM 2007; GON 2011a, b; MOEF 2005, 2012a, b; GOIRP 2003). However, water-related hazards and extreme weather events have been identified as the recurring cause that trigger difficulties of disastrous proportions in those countries, which eventually causes crop and income loss for the poor towards setting up food insecurity (GTZ 2010; GIZ undated; Thapa et al. 2010; ROM 2007; Bhatti 1997; Leichenko and Wescoat 1992; GON 2011a, b; IWMI 2009; Eriyagama et al. 2010; Ahmed 2005; Alam et al. 1998; Sharma and Sharma 2008). Water therefore is thought to be the key element to cause food insecurity under climate change. The following table (Table 7.1) provides a bird's eye view of stated objectives concerning climate change adaptation by various South Asian Countries.

Almost all South Asian countries have already developed country-specific climate change strategies and/or action plans. Each country discusses adaptation needs based on future vulnerabilities, though time frames often do not match with each other, as has been indicated in the above table. For example, the Bangladesh Climate Change Strategy and Action Plan (BCCSAP) has set out a timeline until 2015,

Table 7.1 Country-specific CCA objectives, as stated in nationally available documents

Name of country	CCA objective(s), as stated in national documents	Timeline/ Interim timeline
Bangladesh	To ensure food security (MOEF 2009)	2010–2015
Bhutan	To promote adaptation in agriculture and hydropower sectors as the two most vulnerable sectors in Bhutan to climate change To create awareness and develop adequate capacity to respond to future threats of climate change (RGOB undated)	No specific timeline is mentioned
India	To enhance adaptive and mitigative capacity of the economy To promote vigorous infrastructure growth, keeping in mind the long-term perspectives of ensuring lesser impacts due to likely climate change (MOEF 2012a, b)	No specific timeline is mentioned
Maldives	To assist adaptation in priority sectors: (1) land, beach and human settlement; (2) critical infrastructure; (3) tourism; (4) fisheries; (5) human health; (6) water resources; (7) agriculture and food security; and (8) coral reef and biodiversity (ROM 2001)	2020
Nepal	'Under Nepal's NAPA, the strategy and actions have been targeted to increase community's adaptive capacity …' through various means (GON 2004). One of the approaches considered in NAPA is a 'watershed and landscape level approach dealing with issues related to food security…' (GON 2004)	No specific timeline is mentioned
Pakistan	To ensure water, food and energy security To minimize the impacts of natural disasters on human life, health and property (GOIRP 2003)	2030
Sri Lanka	'To increase Sri Lanka's resilience to climate change impacts' in various sectors, which include (a) agriculture and fisheries, and (b) water resources (Anonymous 2010)	2011–2016

where food security is given due attention (MOEF 2009). In contrast, the Government of India (GOI) announced the preparation of India's Action Plan on Climate Change, set up the National Water Mission out of the eight missions to deliver adaptation in water resource sector by 2017 (GOI 2009). Most of the timelines in South Asian short-term adaptation targets are set between 2015 and 2030, because the model-based vulnerability assessments have timelines between 2020s and 2030s.

The issues concerning adaptation in water sector not only connect food security, but it also highlights protection against water-related invigorated hazards and disasters (MOEF 2009, 2012a, b; INCCA 2010; Bhatti 1997). Protection of population from floods, droughts and cyclonic storm surges are the key concerns which revolve around adaptation discourse involving water resources (Ahmed 2005; Chitranarayana and Punyawardana 2008; MOEF 2009, 2012a, b; Ahmad et al. 2003; Nishat and Mukherjee 2013). The available government documents emphasize on protective infrastructure (for example, to safeguard standing crops), irrigation and water efficiency, coastal protection infrastructure and drainage infrastructure. Although it might appear that the adaptation modalities are too biased towards engineering-based solutions, the inherent objective is to protect the interests of the affected people—in most cases the farming communities in respective South Asian countries. In most cases these infrastructural adaptations are thought to be put in place with respective Government's finance so that the poor beneficiaries do not have to pay directly for their adaptation services (Thapa et al. 2010; Dutta et al. 2013; Oxfam 2011; GOIRP 2003; MOEF 2009). In addition to Government's efforts, South Asian NGOs have been working hard towards enhancing adaptive capacity of the poor and adversely affected people by means of community-based adaptations (CBA)—while the approach itself has been developed in South Asia and tested in many different conditions throughout the region.

There is no denying the fact that many of the adaptation measures in water sector are not new, rather their respective efficacies are well tested and understood in the region. In most cases, the existing measures will be replicated where needed and/or the old design will be realigned with the necessity which has been brought to surface by exacerbated impacts of climate change. As it has been planned, the old structures will be refurbished to suit to anticipated/projected situations so that climate change cannot make such responses redundant. Efforts are being made to establish synergies between known disaster risk reduction (DRR) approaches and climate change adaptation (CCA) approaches (GTZ 2010; MOEF 2009, 2012a, b; Nishat and Mukherjee 2013).

7.3 Trends in Adaptation in South Asia

All the seven South Asian countries are developing countries, while Bangladesh, Bhutan and Nepal are struggling with the status of Least Developed Country (LDC). Until the first decade of the present century, none of the South Asian countries has started adaptation activities in a major way, excepting a few trials on CBA. In the early years of adaptation, South Asian countries have just started to organize themselves towards designing and implementing adaptation.

In Nepal and Bhutan, GLOF bypasses are being considered to avoid risks of GLOF occurrence, which might not have been considered without climate change. These initiatives may be termed as stand-alone CCA activities. Most of the stand-alone CCA activities are preparatory in nature and/or concentrating on research. Conducting field research and subsequent extension of hazard resistant crops are activities which will provide the farmers with better options to deal with climate change. No immediate results may be found and farmers may not be willing to experiment with such new varieties; however, such research and their extension appear to be important for widening the resource pool to address new climate thresholds in crop production.

While activities with targeted CCA objectives are few, majority of the DRR and sectoral development projects in the South Asian Region account for adaptation co-benefits. Climate change has certainly given new impetus to revitalize early warning systems for water-related extreme events, which are likely to remain effective under a warmer world. Drainage infrastructures in rural roads as well as regional and national highways are now being designed, keeping a greater draft to facilitate drainage of higher run-off volume (for example, in Bangladesh, MOEF 2012a, b). Throughout South Asia, new food silos are built to maintain a healthy stock of food, only to be utilized during food insecurity. Efforts are now made to build a regional food bank, where food will be mobilized from such silos whenever food shortage is sensed in any of the regional countries.

Most of the CBA activities are perceived to enhance adaptive capacity of communities, which are expected to facilitate coping of recipient households and communities with or without climate change.

7.3.1 Policy, Strategy, Legal Provisions and Governance

South Asian countries are in the process of developing country-specific strategies, policies and action plans (GOI 2009; MOEF 2009; GON 2011a, b; Anonymous 2010). While Bangladesh, India and Sri Lanka have developed respective strategies and action plans towards implementing CCA, Nepal developed its Local Adaptation Plans of Action (LAPA) (MOE 2011a, b). In the backdrop of a general lack of financing for CCA, only a handful of CCA projects are being implemented in South Asia. Policies only to promote CCA are yet to be devised in many of the countries; however, certain policy elements are found in various statements and country-specific development discourses.

While greater efforts towards establishing practices of good governance are exhibited in policy formulation processes, it is found that similar efforts are not particularly exhibited in implementation stages (Oxfam 2014). It is found that in Bangladesh, governance is not adequately designed to integrate concerns of stakeholders, rather it follows interests of political masters and bureaucrats, often disregarding people's views (Alam et al. 2011; TAF 2012). Similar experiences are reported in case of India and Nepal (Dutta 2013a, b; Fujiwara 2010; Pant and Gautam 2013; Dhungana et al. 2013; Regmi and Bhandari 2013).

South Asia is somewhat blessed in terms of political integration of climate change. However, the absence of policies on CCA often diverts resources to business as usual development and CCA concerns are relegated in decision-making processes. Especially in the two LDCs, political integration towards CCA implementation is found to be rather poor (TAF 2012; Ghimire 2011). One of the issues is lack of legal provisioning, which is found to be true across the region. South Asian nations are deeply engaged in international climate negotiations; however, the respective political masters are too reluctant to frame laws and regulations to drive CCA in their own countries. Part of the problem lies with lack of resources to cater the growing adaptation needs of the poor, which often discourages the political masters to commit to something tangible. Moreover, international financing on CCA has not been materialized yet under the UNFCCC processes. CCA should ideally be placed under proper legal framework in South Asian countries—which ought to be done irrespective of achieving an equitable and legally binding climate regime through the negotiations.

7.3.2 *Institutional Arrangement*

South Asian countries apparently are struggling to come up with strong institutional arrangement towards implementing CCA (Alam et al. 2011; Chhetri et al. 2011; Davis and Li 2013; Dutta et al. 2013). In cases of Bhutan and Maldives, perhaps because of their respective sizes compared to bigger South Asian neighbours, the institutional arrangements are relatively better defined. On the other extreme, the same for India is also better organized, though there still are criticisms (Dutta 2013a, b). The LDCs are struggling to define acceptable institutional framework which would streamline CCA activities across sectors and institutions and ensure better coordination (Alam et al. 2011; TAF 2012; Chhetri et al. 2011). For each of the South Asian countries, there is a focal institution. In most cases, the issue is being dealt with Ministries/Commissions in charge of Environment. However, CCA must be coordinated well with other ministries, particularly with ministries in charge of water resources, agriculture and food security, disaster management, infrastructure, human health, women's affairs, social development, planning and finance. Effective coordination mechanisms are yet to be developed.

Bangladesh has been the first South Asian country to provide for own finance through budgetary allocation since 2009–2010 fiscal year. While the legal provisions are largely adhered to across the region and beyond, very limited mechanisms have been put in place to ensure (a) good governance through appropriate monitoring and evaluation frameworks and incorporating feedbacks in the project cycle and (b) coordinated CCA spending and for ensuring good governance through monitoring, evaluating and capturing feedbacks in the project cycle (Alam et al. 2011). The initial enthusiasm and efforts are eventually proved to be not so productive, which forced the country to gradually decrease budgetary allocations in subsequent fiscal years, despite a six-fold increase in overall budgetary allocation in about 7 years (i.e. between 2009–2010 and 2015–2016 fiscal years).

Climate finance governance is also found to be flawed in the Maldives (Naeem 2013), India (Dutta 2013a, b) and Nepal (Regmi and Bhandari 2013). While in India allegations have been made to allocate money for CCA projects where adequate scientific justifications could not be presented (Dutta 2013a, b), in Nepal projects which have been received relatively lesser priority by the stakeholders have been financed as priority projects (Regmi and Bhandari 2013). In Maldives, financing decision-making have highlighted economic objectives rather than human vulnerabilities (Naeem 2013). In most cases in South Asia, political economy driven decisions dictated terms in financing projects with an objective of adaptation. Scientifically just, people-centric and gender sensitive projects on CCA are yet to be given adequate importance in the decision-making processes.

7.3.3 *Integration and Mainstreaming*

In the formative years of climate action (i.e. between 1995 and 2010), CCA was perceived hastily. The LDCs were asked to develop National Adaptation Programme for Action (NAPA), which yielded a few pilot scale project ideas in Bangladesh, Bhutan and Nepal (MOEF 2005; ROM 2007; GON 2010). Even in Maldives, bilaterally funded CCA projects have been perceived and implemented without bringing synergies with national planning processes (Naeem 2013). Much less emphasis was given on planned approach and integration with sector-specific development processes and plans.

Only recently, greater efforts are considered towards integration of CCA with DRR and development. Rather than stand-alone CCA projects, CCA activities are increasingly perceived as incremental to business as usual development projects (GED 2014 in Bangladesh; GOP 2012; GOI 2009; Anonymous 2010). While Bangladesh, Pakistan and Sri Lanka emphasized on strategic moves towards integration with various sectors, India committed to eight national missions with a view to integrate CCA with various relevant development sectors (MOEF 2009; GOP 2012; Anonymous 2010; GOI 2009).

The Article 4 of UNFCCC calls for a collaborative adaptation regime where the Annex-1 Country parties are supposed to assist developing vulnerable countries, including LDCs and SIDS, to implement adaptation (United Nations 1992). The provisions have been integral part of Bali Action Plan and the negotiated text of Ad-hoc Working Group on Long-Term Cooperative Agreement (AWG-LCA) as well as the Ad-hoc Working Group on the Durban Platform for Enhanced Actions (ADP). Meanwhile, the formation of the Green Climate Fund (GCF) has created provisions for financing adaptation. Therefore, the South Asian countries have been anticipating UNFCCC-led financing, especially from GCF towards implementing CCA activities. The opportunity for international financing for adaptation has in turn dampened local provisioning of finance for adaptation in South Asia, as in the rest of the developing world.

Although the need for CCA is well established in official documents of South Asian countries, little has so far been done to change design standards, especially in designing water infrastructures. Provisioning for climate change has been remained arbitrary, often left to the intuition of engineers involved in particular projects. In many cases, climate change related hard data is not collected or analysed, which might be a prerequisite to change design criteria. For example, the coastal protective embankments in Bangladesh are refurbished, where the crest height is adjusted by 30 cm to cease overtopping by invigorated tidal waves (Nishat 2015). The value has been considered arbitrarily in the absence of any hard data on sea level rise. Adaptation is surely going on, however, in the vacuum of any scientific design standard.

7.3.4 Finance

As indicated earlier, there has been little provisioning of finance—both internally and externally. National and sub-national level financing for CCA had not gained momentum perhaps, because of the political constituencies were waiting for the UNFCCC process-led financing to come through. Since the entire region is more or less fighting against abject poverty and food insecurity, much of the regional financing go for addressing such sectors, in addition to human resource development (education, health) and much needed infrastructure. Climate change has not been considered high in the development agenda, which is also causing a slow start in internal financing for CCA. The Climate Policy of Pakistan (GOP 2012) focused primarily on external financing, giving little consideration for internal financing for CCA.

International financing for CCA is likely to come through the formation and operationalization of the GCF. GCF is exhibiting early readiness for climate financing with its US$ 10.4 billion fund, part of which will be dedicated for CCA projects. Despite the fact that the Annex-1 parties have claimed that they had mobilized US$ 33 billion as climate change financing, much of the said amount had been mobilized as Overseas Development Assistance, while the actual climate financing is estimated to be in the order of only 10% of the total amount. This shows the scanty amount being mobilized so far in global climate financing, let alone CCA financing for South Asia.

There is no denying the fact that there have been bilateral donor interests in early years to finance CCA in South Asia. In 2009, 18 donors formed a financing group to assist Nepal with CCA financing. Similarly, the other LDC Bangladesh formed a Bangladesh Climate Change Resilience Fund (BCCRF) with donor financing on climate change (Alam et al. 2011). However, bilateral donor support did not come through in a major way. Rather, both the LDCs have been brought under the PPCR programme, where loan financing has been offered to address climate change. Both Nepal and Bangladesh have been using such loans (Alam et al. 2011; CCNN 2011), in the backdrop of diminishing international support on CCA through bilateral financing.

There is no denying the fact that the urgency for implementing CCA activities in South Asia has not been translated into firm actions due primarily to lack of financing. The GCF is about to begin financing CCA projects starting July 2015, in line with its pre-2020 commitments. Although none of the countries could complete accreditation of their respective National Implementing Entities (NIE) to access finance directly from the GCF (as of June 2015), the South Asian countries still can access financing from the GCF through Multilateral Implementing Entities (MIEs). It remains to be seen how regional countries can make best use of the GCF financing for addressing CCA.

7.3.5 Implementation Capacity

There have been mixed experiences in implementation of early CCA activities in the region. South Asia is not a homogenous region in terms of institutional and human capacities. Quality of implementation in India and Sri Lanka appears to be far better than in Bangladesh and Nepal. The available national documents on climate change often highlight needs for enhancing national capacities (Meenawar and Sovacool 2011; RGOB 2005; Dutta et al. 2013; Regmi and Bhandari 2013; Parry et al. 2013; MOE 2011a, b; Ministry of Environment and Natural Resources 2007; MOEF 2012a, b). However, adequate efforts are not being taken to address such lack of capacities, especially those that involve institutional and human capacities to implement projects judiciously.

The analysts have identified such lacking as critical for Bangladesh, Nepal and Pakistan, which might jeopardize future CCA activities (TAF 2012; Regmi and Bhandari 2013; Parry et al. 2013). Other than in India and Sri Lanka, few resources have been dedicated so far to enhance human capacities. The absorption capacity of the respective lead institutions (e.g. Ministry of Environment and Forest for Bangladesh, Ministry of Science, Technology and Environment for Nepal and so on) is found to be rather poor, especially in Bangladesh and Nepal (TAF 2012; Chhetri et al. 2011). In contrast, Maldives and Bhutan have only a small national pool of officials. It is easy to enhance capacity of the small pool of officials. Unless the pools are duly expanded and responsibilities are designated, it would be a gigantic task to deliver all CCA activities—even if international finances are made available to such countries.

India has a proven edge in South Asia regarding absorption capacity. Not only that it has the largest population, its human capital is well equipped to tackle the tasks ahead of them. India has proven its institutional capacity by becoming the second largest implementer of CDM projects under UNFCCC processes. Yet, India wishes to enhance its institutional and human capacity further in order to address the issues concerning climate change.

7.3.6 Technology

Once again, there are certain countries in South Asia such as India and Pakistan which are ahead of others in terms of technology innovation, development and adoption in relation to CCA. However, unlike mitigation technologies, adaptation technologies have generally been introduced and tested in country-specific situations and the countries in the region are not starting from zero. Therefore, some capacities already exist in the regional countries. If these capacities are utilized properly with proper scientific guidance, analysis and resulting design criteria, technologies may well be made use of if those are accessed by the regional countries.

Accessing technology largely depends on willingness and ability to do so by the respective countries. Regionally, much emphasis has been given on developing and adopting computer aided technologies on generation and subsequent dissemination of early warning systems (with respect to floods, flash floods, cyclones, drought advisories, etc.) and protection infrastructures.

7.3.7 Decision-Making Tool and Institutionalization

All the South Asian countries have produced vulnerability assessment reports, though their analytical rigour and output qualities largely vary. The smaller countries mostly depend on external support (technical as well as financial supports) to complete vulnerability assessments (UNDP undated; Marambe 2014). While major sectors are covered in available vulnerability assessments, those are completed with gross approximations and using assumptions, often in the absence of validated climate modelling results. In most cases concocted 'what if' scenarios are used, often disregarding dynamic nature of other (i.e. non-climatic) vulnerability contexts. In almost all the countries, proper economic analysis of sector-specific vulnerabilities is not done yet. There have been little efforts to apply analytical hierarchy methods towards development and application of decision support tools.

Other than in India, very little analysis has been done to review vulnerability assessments. There is hardly any effort to reveal adaptation gaps in respective countries and sectors. Countries have not developed yet any noticeable set-ups to periodically carry out analysis and provide updated information on vulnerability. As if without such rigorous analysis, respective political masters will allocate resources to invest in CCA!

7.3.8 Awareness and Access to Information

The most unfortunate part of the CCA discourse in the region deals with lack of general awareness and limited access to information. Local people in any of the countries can relate to changing climatic parameters, citing their own experiences (NCVST 2009; MOEF 2005; Dutta et al. 2013). However, the vulnerability and sectoral impacts, although assessed and duly published, have not been disseminated in forms and languages which are palatable to the local mass in almost all the South Asian countries. An analytical report by the Asia Foundation (TAF 2012) clearly criticizes that the well-revered BCCSAP in Bangladesh has not been shared with large number of stakeholders in any tier below the national level. LAPA is another case, where some local people have been mobilized to take part in the development of local level CCA plans; however, larger stakeholders are yet to be reached with the outcomes (Ghimire 2011). National media, in most cases, have been kept largely outside the purview of dissemination or they did not pay adequate attention to impacts of and adaptation to climate change.

People have been responding to cope with changing trends, however relying mostly on their ancestral guidelines and known low-cost technologies (Dhungana et al. 2013; Ahmed 2010; MOEF 2012a, b; Chapagain et al. 2009). Many people in South Asia have been using cell technologies for communication. Despite the fact that a large proportion of them are still living in abject poverty, the relative rate of adoption of cellular technology has been phenomenal throughout South Asia. And yet, early warning systems across the region have failed to make best use of such important technologies. This gives a flavour how technologies are being thought as medium to disseminate information in relation to climate induced risks and adaptation.

Local authorities have not been properly integrated (with an exception in India where Village Panchayats are somewhat engaged in managing floods, etc.) on issues relating to CCA (TAF 2012; Alam et al. 2011; Dhungana et al. 2013; Regmi and Bhandari 2013). Lack of monetary flow and information have created disconnect between central government and local authorities, although it is well understood that without their involvement, CCA implementation might never be fully realized.

7.4 Current Gaps and Areas of Adjustment

From the above analysis, it is evident that there are significant gaps towards achieving 'maintenance adaptation'. Since CO_2 concentration in the atmosphere has surpassed 400 ppmv levels, the world is projected to face a rise in surface temperature exceeding 3.8 °C by 2100 with respect to 1990 levels. Therefore, an 'enhanced adaptation' regime is warranted to address the rising temperature and its adverse implications. In this context, the gaps in CCA actions appear to be widening if one considers needs for enhanced adaptation and the current level of CCA in South Asia.

The following table (Table 7.2) summarizes the current gaps in relation to maintenance adaptation for the eight indicators explained in the framework chapter (Table 7.3).

Table 7.2 Qualitative analysis of prevailing gaps in adaptation considering 'maintenance adaptation' in South Asia

		Effectiveness in formulation and implementation			
Indicators	Adequacy	Presence/absence of instruments and practices	Quality of overall delivery	Level of compliance	Comments
Policy, strategy, legal provision and governance	Independent policies are emerging No prioritization of CCA projects There is no legal provision as yet	Structures are vaguely present, emerging	Not satisfactory Not being pro-people yet	There are inherent weaknesses Lack of legal provision has been adversely affecting compliance mechanism	Although the region is lagging behind in terms of policies, legal provision and governance, there is increasing willingness amongst regional nations to improve upon in these areas
Institutional arrangement	Lack of coordination Institutional arrangement emerging	Instruments and practices are present, however remain weak	Poorly arranged	Poorly guided and coordinated	Most of the countries have weaknesses in institutional arrangements. Current emphasis is on mainstreaming through inter-agency coordination
Integration and mainstreaming	Lack of understanding in different tiers Dearth of data to facilitate integration	Absent	Poorly integrated	Poorly managed	Moving towards integration and mainstreaming (viz., India, Bangladesh, Pakistan, Sri Lanka)
Finance	Almost non-existing (compared to set targets)	Largely absent, even if present, largely inadequate	Inadequate (poorly mobilized in Nepal and Bangladesh)	Monitoring and evaluation is poorly practised or even bypassed (viz., Bangladesh)	Almost all the countries have been facing financial provisioning, from both internal and external sources. Bilateral sources are yet to make any significant marks in CCA financing

Implementation capacity	Inadequate capacity Major knowledge gap exists in tiers below central level Lack in M&E capacity No review mechanism has been established	Instruments and practices are present, however remain weak	Major gaps exist (with an exception of India perhaps)	Inadequate	With exceptions in cases of India and perhaps Sri Lanka, implementation capacity is rather weak throughout South Asia. However, noticeable efforts are being made to improve upon existing capacities
Technology (development, utilization, capacity)	There has been some development Lack of integration with end users	There are instruments and practices with respect to technologies	Improving, however innovation is needed Inherent weaknesses in dissemination modalities	Emerging	Much pronounced efforts are needed on R&D; gender sensitivity needs to be plugged in while developing and promoting new technologies
Decision-making tools and institutionalization	Not developed A democratic decision-making has not been practised There is no prioritization of CCA actions	Largely absent	Poorly delivered (political economy dictates terms, largely depends on wishes of power holders)	Not existing Non-emerging!	Democratic norms must be developed and practised
Awareness, knowledge and access to information	Growing Large gaps existing	Present, though weakly perceived	Poor Often not sensitive to local contexts	Poor	General awareness is still not up to the mark. Must be integrated with education system Media must take lead in raising mass awareness

Table 7.3 The key shared challenges in South Asian countries towards delivering CCA

Assessment Indicators	Common challenges faced by South Asian countries towards delivering CCA
Policy, strategy, legal provisioning and governance	Policy harmonization involving other sector-specific development policies hasn't taken place Good governance in all stages of CCA involving project selection to executing project exit plan has been inadequate Legal provisioning has remained a major gap
Institutional arrangement	Proper institutional arrangement, including coordination mechanism, supervision, monitoring and oversight roles in relation to CCA activities are not adequately developed
Integration and mainstreaming	Clear national vision towards integration and mainstreaming has not yet emerged, although a few countries have boldly accepted the policy of mainstreaming CCA into development
Finance	All the countries have been struggling with inadequate climate financing and waiting for international 'new and additional' financing to come and boost up CCA activities
Implementation capacity	The current capacity in most cases has been below required level, which needs to be enhanced rather quickly to facilitate implementation of CCA activities
Technology	There exist challenges in terms of CCA technology development, adoption at all necessary tiers and inter-regional cooperation, where applicable
Decision-making tool and institutionalization	There are weaknesses in using decision-making tools, which makes room for decision makers to arrive at a CCA decision which is based on political economy
Awareness and access to information	Public awareness is low, which needs to be increased in order for general mass to be integrated in participatory decision-making processes

7.5 Voluntary Pledges Under Nationally Determined Contributions

Nationally Determined Contributions (NDCs) is the outline of post-2020 climate actions that the Parties to the Convention intend to take under a new international agreement, set to be signed in the COP21 in Paris, France. As per the COP decision 1/CP.19, the country Parties are requested to initiate or intensify domestic preparations towards achieving the objectives of the Convention as set out in its Article 2, without prejudice to the legal nature of the contributions. Such publicly made expression of intentions, by definition, is thought to be 'nationally determined' and 'contributory', while referring primarily to country Parties' ambition to contribute to mitigation (i.e. Article 2). South Asian countries have developed their own NDCs and communicated duly to the UNFCCC Secretariat (GOI 2015; GON 2016; GOP 2015; KB 2015; MMDE 2015; MOEE 2015; MOEF 2015). A few regional NDCs have also touched upon their contributions to adaptation actions, including

Table 7.4 Summary pledges on adaptation made under NDCs of South Asian countries

Name of country	Adaptation priorities	Cost of adaptation	Status of implementation	Timeline/Interim timeline
Bangladesh	Food security, livelihood and health protection (incl. water security); comprehensive disaster management; coastal zone management including control of salinity intrusion; flood control and erosion protection; building climate resilient infrastructure; increased rural electrification; enhanced urban resilience; ecosystem-based adaptation (including forestry co-management); community-based conservation of wetlands and coastal areas; policy and institutional capacity building	Bangladesh may experience an annual 2% loss of GDP by 2050 because of climate change Over the last three decades, the government of Bangladesh has invested over $10 billion (at constant 2007 prices) to make the country more climate resilient and less vulnerable to disasters induced by climate variability and change No indication for future adaptation cost requirement	Total allocation to development activities amounts to about 32% of the national budget	Near term plans and actions (no specific time frame)
Bhutan	Water security through integrated water resource management (IWRM) approaches; promote climate resilient agriculture; sustainable forest management and conservation of biodiversity; strengthen resilience to climate change induced hazards; minimize climate-related health risks; climate proof transport infrastructure against landslides and flash floods; promote climate resilient livestock farming practices; enhancing climate information services for vulnerability and adaptation assessment and planning; promote clean renewable and climate resilient energy generation; integrate climate resilient and low emission strategies in urban and rural settlements	The scale of funding available to address both development needs and the additional burden of mitigation and adaptation will be significantly higher than presently available No specific cost estimation has been highlighted	Bhutan is already spending its own resources for some climate change adaptation and mitigation actions through the budgeting for the current 5-year plan	NDC will be considered and integrated in the preparation of the 12th 5-year development plan (2018–2023) and also subsequent 5-year plan periods

(continued)

Table 7.4 (continued)

Name of country	Adaptation priorities	Cost of adaptation	Status of implementation	Timeline/Interim timeline
India	Out of the eight national missions on climate change, five missions focus on adaptation in sectors such as agriculture, water, Himalayan ecosystems, forestry, capacity building and knowledge management. Climate plans at the subnational level also focus significantly on adaptation	India's expenditure on programmes with critical adaptation components has increased from 1.45% of GDP in 2000–2001 to 2.82% during 2009–2010 Expenditure on human capabilities and livelihoods viz., poverty alleviation, health improvement and disease control and risk management constitutes more than 80% of the total expenditure on adaptation in India No specific cost estimation or future requirement is cited	32 states and union territories have put in place the state action plan on climate change (SAPCC), attempting to mainstream climate change concerns in their planning process. Key sectors covered by SAPCCs include agriculture, water, habitat, forestry, health and disaster management amongst others	No specific timeframe is mentioned
Maldives	Maldives aims to undertake adaptation actions and opportunities and build climate resilient infrastructure to address the current and future impacts of climate change. Key areas for adaptation are: Enhancing food security; infrastructure resilience; public health; enhancing water security; coastal protection; safeguarding coral reef and its biodiversity; tourism; fisheries; early warning and systematic observation	Domestic budgetary spending on addressing climate change remains an additional burden towards the achievement of sustainable development. International support is necessary	Public finance is being allocated to meet urgent and immediate adaptation actions	No specific timeframe mentioned

Nepal	Nepal's pilot program for climate resilience (PPCR) compromises four components: i) building climate resilience of watersheds in mountain eco-regions, ii) building resilience to climate-related hazards, iii) mainstreaming climate change risk management in development and iv) building climate resilient communities through participation of private sector	The 2013 study on economic impact assessment of climate change in key sectors (agriculture, hydropower and water-induced disasters) has estimated direct cost of current climate variability and extreme events equivalent to 1.5 to 2% of current GDP/year (approximately USD 270–360 million/year in 2013 prices) and much higher in extreme years Overall, the economic costs of climate change in Nepal for these three sectors could be equivalent to 2–3% of current GDP/year by mid-century	Nepal prepared NAPA in 2010. In 2015, Nepal launched a process to formulate and implement National Adaptation Plan (NAP) to address medium- and long-term adaptation needs Nepal has adopted a National Framework on local adaptation plans for action (LAPA) to ensure integration of adaptation and resilience into local to national planning processes	No specific timeframe mentioned
Pakistan	No key sector has been identified. Adaptation to impacts of climate change is a vast area of untapped opportunities in Pakistan due to its multi-sectoral nature of economy, huge infrastructure needs, distinct climatic zones, ecological systems and administrative arrangements	The investment costs for adaptation interventions are being determined in consultation with the provinces and other stakeholders, and will be conveyed in due course of time	No statement	No specific timeframe mentioned

(continued)

Table 7.4 (continued)

Name of country	Adaptation priorities	Cost of adaptation	Status of implementation	Timeline/Interim timeline
Sri Lanka	Adaptation will focus on key sectors those involving food security, health, water, coastal and marine resources, biodiversity and ecosystems, infrastructure and human settlements	Based on the NCCAS and the NAP, the total finance requirement for adaptation actions for the period of 2016–2025 is estimated to be US$420 million	Sri Lanka intends to contribute to the global adaptation goal by enhancing local climate change adaptation The Sri Lankan government is willing to contribute its finances to achieve the target; however, the level of ambition will always be high with supported actions	Base year: 2010; target period: 2021–2030

indicative estimates of costs associated with such contributory adaptation actions. Table 7.4 highlights such actions in reference to each South Asian country.

As indicated above, the primary focus of country-specific NDCs has a common emphasis on voluntary mitigation actions, not really on adaptation options. However, most of the South Asian countries have generally exhibited strong national support to undertake adaptation actions and remained just indicative to contributory actions on adaptation. Through the NDCs, adaptation priorities are well communicated for South Asian countries. With an exception of Sri Lanka, South Asian countries have not mentioned any specific timeframe under which such strong commitment to adaptation would be fulfilled.

7.6 South Asia's Readiness for Enhanced Adaptation Actions

From the above assessment, it may be inferred that the South Asia region has not been addressing CCA in the right way and the current efforts are falling short of business as usual or maintenance adaptation. Since a worse climate regime might be unavoidable, especially if the emission peaking is delayed and current large emitters do not commit and legally binding targets are not agreed upon, the current level of gaps will certainly be widening with respect to a hypothetical 'safe' level of adaptation.

The policy gaps need to be addressed immediately and inter-sectoral synergies must be established in order to optimize benefits from policy integration. CCA should not be seen as a unimodal mechanism. A pronounced legal provisioning should also be considered where respective constitutions should make rooms for an integrated approach to deliver CCA.

In the backdrop of rather large gaps in policy synergy and legal provisioning, the current institutional arrangement would not enable regional countries to deal with climate change involving enhanced CCA. Before entering into post-2020 climate regime, the regional countries must take immediate actions to enhance their respective institutional capacities. The coordination efforts must be given high priority, otherwise the regional countries will fall short in terms of readiness to deliver enhanced adaptation actions.

The integration and mainstreaming efforts offer golden lines amidst the dark clouds. Apparently, most of the South Asian countries have started to realize that CCA needs to be integrated with development and accordingly, CCA should be mainstreamed with planned development pathways. Both India and Bangladesh have been shaping up their development actions by integrating climate change, where both adaptation and mitigation are taken into account.

Finance remains the major stumbling block towards delivering CCA actions in South Asia. While greater efforts must be made to facilitate integration of CCA into development, as far as possible with internal financing, the respective countries must align its institutions to take advantage of the emerging financing regime through the operationalization of GCF. GCF worthy CCA concepts need to be developed into full proposals so that projects are conceptualized regarding adaptation, however keeping social and environmental norms at the core of such concepts.

It does not matter whether there will be internal or external financing for CCA activities, greater efforts must be made to address the existing gap regarding implementation capacity. The analytical as well as M&E capacities need to be enhanced as quickly as possible in order to exhibit readiness to take advantage of emerging global opportunities on climate change. The current implementation capacities are too shallow to justice to the cause of CCA.

While decision-making should be based on right data and analysis, prioritization of projects should be done judiciously and backed up by thorough economic, social and environmental analyses, a much greater effort is needed to place science-based knowledge towards determining which CCA project needs to be carried out in which timeframe. Such decision-making processes should dictate which technologies should be used and adopted. Greater care is also needed to choose the right technologies for CCA. While efforts of integration of indigenous knowledge need to be made, much greater efforts are needed to innovate and adopt new and advanced technologies.

Mass people are generally seen as passive recipients of CCA activities, not really as the drivers in this entire pursuit. Without enhancing people's awareness and knowledge, the objective of CCA activities may not be fully achieved, even if the financing jinx is somehow overcome. The current role of academia is found to be rather weakly perceived. Media is not keen on highlighting the need for addressing climate change. The media should be brought in towards informing the mass population, a significant proportion of which is still illiterate. Currently, no effort has been made to integrate the topic in the curricula.

7.7 Attaining Adaptation Vision in South Asia Region

The rationale and needs for attaining adaptation vision in South Asian Countries cannot be overemphasized. And yet, there exists large gaps in perceiving and delivering CCA actions. Not only the current gaps need to be addressed, greater efforts must be put forth to overcome all existing barriers so that a harmonized and well-integrated culture of adaptation prevails and practised throughout the South Asian region.

The focus of CCA in South Asia is found to be in the right place; it involves agriculture and food security, while highlighting the need for integration with water resources sector. That is the common feature of initial adaptation assessments in all parts of the region. This righteous focus must not be allowed to diminish and efforts must be there to plan CCA accordingly and allocate finance so that early adaptations may be initiated without delay.

The poorly organized delivery of CCA in the prevailing lack of coordination amongst actors and institutions needs to be addressed as a priority. Climate change should not be perceived as a mere unimodal action, rather it should create rooms for multiple actors to play their respective roles under a coordinated approach. An institutional arrangement in each of the South Asian countries must facilitate greater coordination and synergy.

The above cannot be achieved without catering the need for enhancing capacity of institutions and their respective human resource pool. Immediate actions are necessary to address the current gap, while an integration of the education system will enable to produce future actors who will have adequate academic understanding to deal with the problem in future. Climate change should be given space in existing curricula throughout South Asia, since the issue is there to be dealt with in future decades to come. The bureaucracy also should be targeted for a skills enhancement programme so that they may participate pro-actively towards integrating CCA into development. Similarly, members of Parliaments of respective South Asian countries should be made aware of climate change so that they may take part in framing legal provisions and oversee the progresses on CCA actions in different sectors.

South Asian countries neither could allocate large amounts to implement adaptation, nor could tap resources from existing funds such as Adaptation Fund, Special Climate Change Fund (SCCF) and Least Developed Country's Fund (LDCF). Other than India, the NIE accreditation processes are found to be too cumbersome for the South Asian countries to get by, while their poor fiduciary management practices appeared to be the barriers towards not being successful in accreditation of respective NIEs. Effort must be made now to carefully analyse the situation in order to become ready for the GCF. The pre-2020 window is about to be operationalized soon. Due to lack of readiness, South Asian countries might have to deal with MIEs instead of NIEs, which is generally acknowledged as counterproductive to the spirit of 'direct access' for the developing countries to these funds. Much greater efforts are needed to overcome the existing barriers which involve fiduciary management and compliance with social and environmental norms.

CCA decisions are not generally taken democratically, by putting affected people's concerns and aspirations as central in the designing stage for each CCA action. The current modality of people-centric design of CCA actions needs to be rectified. Adequate norms and practices must be developed so that socially responsive, gender sensitive and environmentally sound CCAs are designed by respective institutions, without being biased to political economy driven decisions.

In the implementation processes, there should be better integration with local government institutions (LGIs). Locally elected leaders and bodies are generally kept in dark when comes to resource allocation for CCA at local levels. Generally, hardly any knowledge is shared by the central government authorities with the LGIs regarding progresses in climate change related policies, legal provisioning, planning, institutional arrangements and financing. Efforts must be made to empower LGI Office Bearers so that they are eligible to take part in planning in local level CCA integration and delivering CCA at the appropriate levels and work in tandem with line agencies and their officials.

References

Agrawala S, Ota T, Ahmed AU et al (2003) Development and Climate Change in Bangladesh: Focus on Coastal Flooding and the Sundarbans. Organisation for Economic Co-Operation and Development (OECD), Paris, 70 p

Ahmad QK, Verghese BG, Iyer R, Pradhan BB, Malla SK (1994) Converting water into wealth: regional cooperation in harnessing the Eastern Himalayan rivers. Academic, Dhaka, p 111

Ahmad S, Bari A, Muhammed A (2003) Climate change and water resources in pakistan: impact, vulnerabilities, coping mechanisms. In: Muhammed A (ed) Climate change and water resources in South Asia. Asianics Agro-Dev International, Islamabad, pp 138–201

Ahmed AU (2005) Adaptation options for managing water related extreme events under climate change regime: Bangladesh Perspectives. In: Mirza MMQ, Ahmad QK (eds) Climate change and water resources in South Asia. Balkema Press, Taylor and Francis Group, Leiden

Ahmed AU (2010) Reducing vulnerability to climate change: the pioneering example of community based adaptation. Care, Dhaka, 154 p

Ahmed AU, Neelormi S (2007) Livelihoods of costal fishermen in peril: in search of early evidence of climate change induced water logging for women in Bangladesh. Centre for Global Change (CGC), Dhaka

Alam M, Huq S (2017) Measuring climate change adaptation: an assessment framework. In: Lee J, Alam M, Sawhney P (eds) Status of climate change adaptation in Asia and the Pacific. Springer, Berlin

Alam M, Nishat A, Siddiqui SM (1998) Water resources vulnerability to climate change with special reference to inundation. In: Huq S, Karim Z, Asaduzzaman M, Mahtab F (eds) Vulnerability and adaptation to climate change for Bangladesh. Kluwer Academic, Dordrecht, pp 21–38

Alam K, Shamsuddoha M, Tanner T, Sultana M, Huq MJ, Kabir SS (2011) The political economy of climate resilient development planning in Bangladesh. IDS Bull 42(3):52–61

Anonymous (2010) National climate change adaptation strategy for Sri Lanka 2011-2016. Ministry of Mahaweli Development and Environment, Battaramulla, p 58

Bajracharya B, Shrestha AB, Rajbhandari L (2007) Glacial Lake outburst floods in the Sagarmatha region: hazard assessment Unsing GIS and hydrodynamic modelling. Mt Res Dev 27:336–344

Bhatti MA (1997) Climate change assessment and adaptation strategies for Pakistan in water sector. Pakistan Science Foundation, Islamabad

CCNN (2011) Governance of climate change adaptation finance in Nepal. Climate Change Network Nepal (CCNN), Kathmandu

Chapagain BK, Subedi R, Paudel NS (2009) Exploring local knowledge of climate change: some reflections. J Forest Livelihood 8(1):106–110

Chhetri N, Chaudhary P, Tiwari RR, Yadav RB (2011) Institutional and technological innovation: understanding agricultural adaptation to climate change in Nepal. Appl Geogr 33:142–150

Chitranarayana RD, Punyawardana BVR (2008) Identification of drought prone agro-ecological regions in Sri Lanka. J Natl Sci Found 36:117–123

Davis M, Li L (2013) Understanding the policy contexts for mainstreaming climate change in Bhutan and Nepal: A Synthesis, Partner Report Series No 10, Regional Climate Change Adaptation Platform for Asia (Adaptation Knowledge Platform). Stockholm Environment Institute, Bangkok, p 20

De Sousa SN, Ahmed AU, Kumar M, Jagtap MD, Sardessai TG, Hassan A (2009) Changes in coastal and marine environments. In: Mitra AP, Sharma C (eds) Global environmental changes in South Asia: a regional perspective. Capital Publishing Company, New Delhi, pp 271–326

De US, Singh GP, Rase DM (2013) Urban flooding in recent decades in the four mega cities of India. J Indian Geophysical Union 17(2):153–165

Dhungana H, Pain A, Khatri D, Gurung N, Ojha H (2013) Climate change and rural institutions in Nepal, DIIS Working Paper 2013. Danish Institute for International Studies (DIIS), Copenhagen, p 16

Dutta S (2013a) Adaptation to climate change in India: a bleak future. In: Dutta S et al (eds) Climate change and India: analysis of political economy and impact. Daanish Books, Gazipur, pp 155–168
Dutta S (2013b) The emerging climate change crisis and the response of the State in India. In: Dutta S et al (eds) Climate change and India: analysis of political economy and impact. Daanish Books, Gazipur, pp 211–234
Dutta S, Ghosh S, Gopalkrishnan S, Bijoy CR, Yasmin H (2013) Climate change and India: analysis of political economy and impact. Daanish Books, Gazipur, p 234
Eriyagama N, Smakhtin V, Chandrapala L, Fernando K (2010) Impacts of climate change on water resources and agriculture in Sri Lanka: a review and preliminary vulnerability mapping. International Water Management Institute (IWMI), Colombo
FAO (2009) The state of food insecurity in the world 2009. Food and Agriculture Organization of the United Nations, Rome
Fujiwara N (2010) The Political Economy of India's Climate Agenda, CESP Working Document No. 325/March 2010. The Centre for European Policy Studies (CESP), Brussels, p 25
GED (2014) Towards Resilient Development: Scope for Mainstreaming Poverty, Environment, Climate Change and Disasters in Development Projects, General Economic Division (GED), Planning Commission, Ministry of Planning. Government of the People's Republic of Bangladesh, Dhaka, p 130
Ghimire S (2011) Climate justice: bottlenecks and opportunities for policy making in Nepal. Discussion paper. ASD-SIAS
GIZ (undated) Adaptation to climate change with a focus on rural areas and India, Deutsche Gesellschaft fur Internationale Zusammenarbeit (GIZ) GmbH, India, New Delhi, p 228
GOI (2009) National water mission under national action plan on climate change. Ministry of Water Resources, Government of India (GOI), New Delhi
GOI (2015) India's intended nationally determined contribution: working towards climate justice, Government of India (GOI), p 38
GOIRP (2003) Pakistan's initial National Communication on climate change. Government of Islamic Republic of Pakistan (GOIRP), Islamabad, p 92
GON (2010) National Adaptation Programme of Action (NAPA) to climate change. Government of Nepal (GON), Kathmandu
GON (2011a) Climate change policy, 2011. Government of Nepal (GON), Kathmandu
GON (2011b) Strategic Programme on climate resilience. Government of Nepal (GON), Kathmandu
GON (2016) Intended Nationally Determined Contributions (INDC). Ministry of Population and Environment, Government of Nepal (GON), p 12
GOP (2012) National climate change policy, Ministry of Climate Change. Government of Pakistan (GOP), Islamabad, p 39
GOP (2015) Pakistan—Intended Nationally Determined Contributions (INDCs). Government of Pakistan (GOP), Islamabad, p 1
GOP/UNEP (1998) Study on climatic change impact assessment and adaptation strategies study for Pakistan. Government of Pakistan and United Nations Environment Program, Islamabad
Goswami BN, Venugopal V, Sengupta D, Madhusoodanan MS, Xavier PK (2006) Increasing trend of extreme rain events over India in a warming environment. Science 314(5804):1442–1445
GTZ (2010) Concept paper: adaptation for climate change through disaster management. Deutsche Gesellschaft für Technische Zusammenarbeit (GIZ), New Delhi
INCCA (2010) Climate change and India: a 4x4 assessment, a sectoral and regional analysis for 2030s. Indian Network for Climate Change Assessment (INCCA), New Delhi
IPCC (2014) The Fifth Assessment Report of IPCC. Intergovernmental Panel on Climate Change (IPCC), Paris
Islam F, Hove H, Parry J (2011) Review of current and planned adaptation action: South Asia. Prepared for the Adaptation Partnership. International Institute for Sustainable Development (IISD), Manitoba

IWMI (2009) Flexible water storage options and adaptation to climate change, IWMI Water policy Brief 31. International Water Management Institute (IWMI), Colombo, p 5
Jianchu X, Shrestha A, Vaidya R, Eriksson M, Hewitt K (2007) The Melting Himalayas: Regional challenges and local impacts of climate change on mountain ecosystems and livelihoods. International Center for Integrated Mountain Development, Kathmandu, 23 p
KB (2015) Intended nationally determined contribution, Kingdom of Bhutan, Communication no. NEC/CC/FCCC/2015/706, Kingdom of Bhutan (KB), p 8
Kelkar U, Bhadwal S (2007) South Asian regional study on climate change impacts and adaptation: implications for human development, human development report 2007/2008, United Nations Development Programme (UNDP), p 47
Kumar KSK, Parikh J (2001) Indian agriculture and climate sensitivity. Glob Environ Chang 11(2):147–154
Lal R et al (eds) (2011) Climate change and food security in South Asia. Springer, Heidelberg
Leichenko RM, Wescoat JL (1992) Complex River Basin Management in a Changing Global Climate: the Indus River Basin in Pakistan. A National Assessment CADSWES. University of Colorado, Boulder, CO
Marambe B (2014) Adaptation planning in Sri Lanka, presentation made in Asia-Pacific Climate Change Adaptation Forum 2014, organized by Asia-Pacific Adaptation Network (APAN), held in KualaLumpur during 1–3 October 2014
Meenawar H, Sovacool BK (2011) Improving adaptive capacity and resilience in Bhutan. Mitig Adapt Strateg Glob Change 16:515–533
MOE (2011a) National framework on local adaptation plans for action (LAPA). Ministry of Environment, Government of Nepal, Kathmandu, Eng version, 20 p
MOE (2011b) Second National Communication on climate change. Ministry of Environment, Democratic Socialist Republic of Sri Lanka, Colombo, p 160
Ministry of Environment and Natural Resources (2007) Capacity assessment and action plan for developing capacity for compliance with global conventions on biodiversity, climate change and land degradation. Ministry of Environment and Natural Resources, Battaramulla, p 164
Mirza MMQ, Ahmed AU (2003) Climate change and water resources in South Asia: vulnerabilities and coping mechanisms. In: Muhammed A (ed) Climate change and water resources in South Asia. Asianics Agro-Dev International, Islamabad, pp 202–228
Mirza MMQ, Ahmed AU (2010) Global warming, changes in hydrological cycle and availability of water in South Asia. In: Global environmental changes in South Asia. Springer, Dordrecht, pp 222–232
Mitra AP (2003) Climate in South Asia and water resources – an introduction. In: Muhammed A (ed) Climate change and water resources in South Asia. Asianics Agro-Dev International, Islamabad, pp 1–20
MMDE (2015) Intended Nationally Determined Contribution, Sri Lanka. Ministry of Mahaweli Development and Environment (MMDE), Battaramulla, p 7
MOEE (2015) Maldive's Intended Nationally Determined Contribution (INDC), Ministry of Environment and Energy (MOEE). Government of Maldives, p 12
MOEF (2005) National Adaptation Programme of Action, Bangladesh. Ministry of Environment and Forest (MOEF), Dhaka
MOEF (2009) Bangladesh climate change strategy and action plan 2009. Ministry of Environment and Forest (MOEF), Government of the People's Republic of Bangladesh (GOB), Dhaka
MOEF (2012a) India second National Communication to the United Nations framework convention on climate change. Ministry of Environment and Forest (MOEF), Government of India, New Delhi, 301 p
MOEF (2012b) Second National Communication. Ministry of Environment and Forest, Government of the People's Republic of Bangladesh, Dhaka
MOEF (2015) Intended Nationally Determined Contribution, Bangladesh. Ministry of Environment and Forest (MOEF), Government of the People's Republic of Bangladesh, Dhaka, p 15

Muhammed A (ed) (2003) Climate change and water resour in South Asia. Asianics Agro-Dev International, Islamabad, 259 p

Naeem AH (2013) In: Shafeega F (ed) As assessment of Climate Finance Governance in Maldives. Transparency Maldives, Male, p 46

NATCOM (2004) India's initial National Communication to the United Nations framework convention on climate change. Ministry of Environment and Forests, Government of India, New Delhi

NCVST (2009) Vulnerability through the eyes of vulnerable: climate change induced uncertainties and Nepal's Development predicaments. Institute for Social and Environmental Transition (ISET), & Institute for Social and Environmental Transition-Nepal (ISET-N), Boulder, CO and Kathmandu

Nishat A (2015) Personal communication with the project consultant. Accessed 16 May 2015

Nishat A, Mukherjee N (2013) Sea level rise and its impacts in coastal areas of Bangladesh. In: Climate change adaptation actions in Bangladesh. Springer, Tokyo, pp 43–50

Oxfam (2011) Minding the money: governance of climate change adaptation finance in Nepal. Oxfam, Kathmandu

Oxfam (2014) Policy Environment in Climate Change Adaptation and Disaster Risk Reduction in Bangladesh. Oxfam Gb, Dhaka (mimeo), p 27

Pant D, Gautam K (2013) Policy provisions and local response on climate change adaptation in Nepal, regional climate change adaptation knowledge platform for Asia, Partner Report Series No. 12. Stockholm Environment Institute, Bangkok Available online at www.asiapacificadapt.net or www.weADAPT.org

Parry J, Keller M, Murphy D (2013) Identifying priority adaptation actions in pakistan: a situation analysis. International Institute for Sustainable Development (IISD), Manitoba, p 33

Pattanaik DR, Rajeevan M (2010) Variability of extreme rainfall events over India during southwest monsoon season. Meteorol Appl 17(1):88–104

Ranger N, Hallegatte S, Bhattacharya S, Bachu M, Priya S, Dhore K, Rafique F, Mathur P, Nville N, Henriet F, Herweijer C, Pohit S, Corfee-Morlot J (2011) An assessment of the potential impact of climate change on flood risk in Mumbai. Clim Chang 104:139–167

Regmi B, Bhandari D (2012) Unripe fruits or non-raining clouds? Climate governance and funding dilemma in Nepal: strengthening governance for mitigation and adaptation to climate change. NAPSIPAG, Kathmandu

Regmi BR, Bhandari D (2013) Climate change adaptation in Nepal: exploring ways to overcome the barriers. J Forest Livelihood 11(1):43–61

RGOB (2005) Bhutan's National capacity self-assessment for global environmental management and action plan, final report. Royal Government of Bhutan (RGOB), Thimpu, 94 p

RGOB (undated) Bhutan National adaptation programme of action. National Environment Commission, Royal Government of Bhutan (RGOB), Thimpu, 73 p

ROM (2001) First National Communication of the Republic of Maldives to the United Nations framework convention on climate change. Ministry of Home Affairs, Housing and Environment, Republic of Maldives (ROM), Male, p 127

ROM (2007) National Adaptation Plan of Action (NAPA): Republic of Maldives. Ministry of Environment, Energy and Water, Republic of Maldives (ROM), Male, p 98

Sharma BR, Sharma D (2008) Impact of climate change on water resources and glacier melt and potential adaptations for Indian agriculture, keynote paper presented at 33rd Indian Agricultural Universities Association vice chancellors' annual convention on 'climate change and its effects on agriculture', held on 4–5 December 2008 at Anand Agricultural University, Anand, India, p 20

Shrestha A, Aryal R (2011) Climate change in Nepal and its impact on Himalayan glaciers. Reg Environ Chang 11:65–77. https://doi.org/10.1007/s10113-010-0174-9

Sivakumar MVK, Stefanski R (2011) Climate change in South Asia. In: Lal R et al (eds) Climate change and food security in South Asia. Springer, Heidelberg, pp 13–30

TAF (2012) A situation analysis of climate change adaptation initiatives in Bangladesh. The Asia Foundation (TAF), Dhaka

Tariq MAUR, van de Giesen N (2012) Floods and flood management in Pakistan. Phys Chem Earth Pts A/B/C 47:11–20

Thapa S, Soussan J, Priya S, Lhendup P, Krawanchid D (2010) Enhancing adaptive capacity in Bhutan and Nepal, Policy Research Brief, No. 1, Regional Climate Change Adaptation Platform for Asia (Adaptation Knowledge Platform), Bangkok, p 9

UNDP (undated) Enhancing the Resilience of the Maldives to climate change through integrating the tourism sector and associated communities into national adaptation. United Nations Development Programme (UNDP), Male, p 5

United Nations (1992) United Nations framework convention on climate change. United Nations, New York

Wang S, Davies RE, Huang W, Gilles R (2011) Pakistan's two-stage monsoon and links with the recent climate change. J Geophys Res (Atmos) 116(D16). Available at onlinelibrary.wiley.com/https://doi.org/10.1029/2011JD015760/full, Accessed 2 Feb 2015

Chapter 8
Status of Climate Change Adaptation in Southeast Asia Region

Ranell Martin M. Dedicatoria and Catherine B. Diomampo

Abstract Like any other region, Southeast Asian Countries are equally vulnerable to climate change and adaptation interventions vary by countries. This chapter examines the status of Climate Change Adaptation (CCA) in SEA by examining present trends and future direction. It has been done through a review of government reports including Intended Nationally Determined Contributions (INDC) to Paris Agreement and related literature. It aims to assess how the sub-region is faring about its adaptation goals and objectives using the eight indicators presented in the framework chapter. It also attempts to identify ways to enhance adaptation actions toward meeting adaptation objectives and goals. Assessment reveals that there remain huge gaps to enhance CCA in the sub-region. These include proper implementation of climate policies and laws, poor coordination among relevant ministries, insufficient and underutilised mechanisms to generate alternative sources of climate financing, economic priorities, limited access to information especially at the level of local communities, and a lack of systems for monitoring and evaluation (M&E) of CCA programmes and projects at the national level. Although it is ideal for countries to aim for enhanced adaptation options to reach their respective adaptation goals, countries would face potential constraints that include cost-effectiveness of adaptation initiatives, the level of uncertainty of risks involved, and the weight of other development priorities, among others.

Keywords Adaptation status · Southeast Asia · Climate change · Adaptation objectives · Mainstreaming · Climate finance · Adaptation technologies

R. M. M. Dedicatoria (✉) · C. B. Diomampo
ICLEI–Local Governments for Sustainability Southeast Asia Secretariat (ICLEI SEAS), Quezon City, Philippines
e-mail: ranell.dedicatoria@iclei.org

M. Alam et al. (eds.), *Status of Climate Change Adaptation in Asia and the Pacific*, Springer Climate, https://doi.org/10.1007/978-3-319-99347-8_8

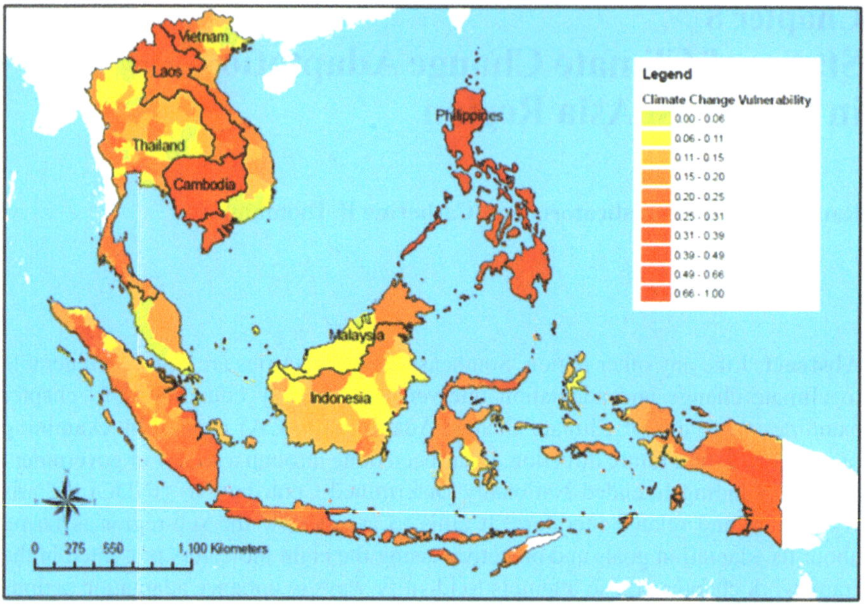

Fig. 8.1 Climate change vulnerability map of Southeast Asia. Source: Yusuf and Francisco (2009)

8.1 Introduction

Considered as one of the world's most dynamic and diverse sub-regions, Southeast Asia (SEA) is composed of 11 countries[1]. It is generally divided into 'mainland' and 'island' zones. The entire sub-region lies between the tropics hence the climate is generally similar—warm temperatures for most parts while upland areas are relatively cool. The sub-region's biophysical features make it highly exposed to hydrometeorological hazards. This, coupled with increasing population and rapidly growing economies with heavy reliance on natural resources, amplifies its vulnerability to climate change.

Yusuf and Francisco (2009) constructed a climate change vulnerability index (Fig. 8.1) for SEA using normalised indicators of exposure (multiple hazard risk exposure), sensitivity (human and ecological), and adaptive capacity. Results showed that most vulnerable areas in SEA include the Mekong River Delta as well as portions of the Philippines primarily due to its frequent exposure to tropical cyclones. While in Cambodia, low adaptive capacities of almost all provinces contributed to its overall vulnerability. Brömmelhörster (2009) explained that climate change resilience ultimately depends on adaptive capacity, which relates to economic, social, and human development.

[1] Brunei, Cambodia, Indonesia, Lao PDR, Malaysia, Myanmar, Philippines, Singapore, Thailand, Timor Leste, and Vietnam. The Association of Southeast Asian Nations (ASEAN), however, does not include Timor Leste, to date.

The UNESCAP (2014) reported that SEA, particularly Indonesia and the Philippines, experienced a total of 527 incidences of natural disasters and 354,293 deaths between 2004 and 2013. While the direct correlation of said disasters to climate change is still a matter of debate, the manifestation of super typhoons, drought, heat waves, sea level rise, salt water intrusion, forest fires among others are reasons enough for the countries to act. With these risks expected to worsen in the future, adaptation is imperative for the countries to mitigate additional losses.

Being signatories to the United Nations Framework Convention on Climate Change (UNFCCC), all 11 countries in the sub-region recognise the relevance of adapting to climate change and have started to act. Responsible national-level entities/bodies have been established, policies and strategies/plans formulated, climate change adaptation (CCA) programmes and projects are being implemented. The influx of climate financing from development banks and aid agencies as well as developed countries has a significant role in aiding SEA countries to overcome high adaptation costs. The sub-region receives a significant amount of financial and technical support to help vulnerable communities to increase their adaptive capacities and resilience. In general, CCA is progressing in the region, but the level of advancement in terms of translating CCA into policy and action varies per country.

The Cancun Adaptation Framework—one of the outcomes during the 16th Conference of Parties (COP16) in 2010—stated that 'enhanced action and international cooperation on adaptation is urgently required to enable and support the implementation of adaptation actions aimed at reducing vulnerability and building resilience in developing countries'. The IPCC fifth Assessment Report released in 2014 reported that there is a build-up of adaptation experiences coming from various regions in both public and private sectors including communities. It further recognised with high confidence the contributions from social (including local and indigenous), institutional, and ecosystem-based adaptation. However, while 'adaptation options exist in all sectors, their context for implementation and potential to reduce climate-related risks differ across sectors and regions'. More recently, the 2015 Paris climate agreement established the global adaptation goals, which include enhancing adaptive capacity, strengthening resilience, and reducing vulnerability to climate change for ensuring an adequate response to the 2 °C temperature goal.

8.2 Adaptation Vision and Priorities

8.2.1 Adaptation Vision

Increasing attention to CCA at the global agenda, acknowledgement of serious and evolving future risks that may impede economic and social development, threaten human security and ecological sustainability, and requirement of more dedicated climate financing for adaptation are all impetus for SEA countries to promote and implement CCA. Being signatories to the UNFCCC, all Southeast Asian countries are aware of and recognise their vulnerabilities to climate change and the need to increase their resilience to be able to better adapt to its future impacts. Through the years, countries in the region have started to incorporate and implement CCA into their policies, programmes, and activities. Except for Brunei, all SEA countries have submitted at least the Initial National Communication in which they report climate change-related activities/studies. In addition, all SEA countries except Timor have submitted their INDCs in time for the 21st Conference of Parties in Paris in 2015. A thorough investigation of the submissions and reports reveals their CCA outlooks and common trends in existing initiatives. Interestingly, developments in advancing CCA vary per country and likely reasons include different levels of vulnerabilities, capacities, priorities, resources, and political will.

In order to assess how CCA is progressing in the region, there is a need to consider how countries envision an ideal CCA and the goals they set for the future. Table 8.1 summarises the adaptation vision/goals/objectives determined by selected SEA countries. Common to the countries' overall goal is to be 'climate resilient' towards achieving sustainable development. Almost all countries with specific adaptation vision/goals recognise the importance of building the resilience of not only livelihoods of vulnerable communities and people, but also vulnerable economic sectors, natural resources, ecosystem, and ecosystem services. Many countries also view CCA as an approach to poverty alleviation, social equity, and improved quality of life. It is worth noting that although Brunei has yet to set its adaptation goal/vision (as it has yet to develop its climate change plan/strategy), the government recognises the importance of climate resilience in protecting its people and their future livelihoods, one of the main goals of Wawasan Brunei 2035 (INDC).

Unlike mitigation targets, adaptation goals/objectives do not have a timeframe and are not quantified. Assessment of whether these objectives and goals have been achieved is very difficult due to inherent characteristics of CCA'uncertainties related to future events, lack of agreement on the definition of 'successful' adaptation, broad range of adaptation options, different timescales, need for counterfactuals and fuzzy cause and effect (attribution)' (APAN-ICLEI 2015).

8.2.2 Priority Sectors for Adaptation

In each SEA country, priority for adaptation depends on the respective country's vulnerabilities to and impacts of climate change on different sectors. Table 8.2 summarises priority sectors for adaptation in the region derived from each country's climate change document (e.g. National Communication; National Adaptation Programmes of Action (NAPA); INDCs, and other climate change policy/strategy/plan of action). The top two emerging CCA priorities in SEA are Water Resources, and Health, which were identified as priority by all SEA countries. These are followed by Agriculture, and Coastal/Marine Resources, which are considered as high priority by almost all SEA countries, except Singapore for agriculture (as it is not an agricultural country) and Lao PDR for coastal/marine resources (as it is a

Table 8.1 Adaptation vision of SEA countries

Country	Adaptation vision/goal/objective	Source
Brunei	One of the principal goals of Wawasan Brunei 2035 is to protect its people and their future livelihoods through enhancing climate resilience and adapting to climate change	INDC (2015)
Cambodia	Vision of Cambodia is to develop towards a greener, low-carbon, climate-resilient, equitable, sustainable, and knowledge-based society. Its goal is to reduce vulnerability to climate change impacts of people, in particular the most vulnerable and critical systems (natural and societal)	Cambodia CC Strategic Plan (2014–2023)
Indonesia	Main objective of Indonesia is the implementation of a sustainable development system which has a high resilience to climate change impacts. Its strategic objectives are to build economic resilience; establish the livelihood (social) resilience to climate change impacts (livelihood resilience); maintain the sustainability of environmental services (ecosystem resilience); and strengthen the resilience in special regions such as urban areas, as well as coastal and small islands Its short-term goal is 'to study and map regional vulnerabilities as the basis of adaptation information system and to strengthen institutional capacity and promulgation of climate change sensitive policies and regulations by 2020' while medium-term goal is to 'reduce risks on all development sectors (agriculture, water, energy security, forestry, maritime and fisheries, health, public service, infrastructure and urban system) by 2030'	National CCA Action Plan (RAN-API) (2013) and INDC (2015)
Lao PDR	Vision of Laos is 'to secure a future where the Laos is capable of mitigating and adapting to changing climatic conditions in a way that promotes sustainable economic development, reduces poverty, protects public health and safety, enhances the quality of Lao PDR's natural environment, and advances the quality of life for all Laotians' Country intends to balance its need for development without compromising its environment. For climate change adaptation, this translates into (a) increase resilience of key economic sectors and natural resources to climate change and its impacts; (b) enhance cooperation, strong alliances, and partnerships with national stakeholders and international partners to achieve national development goals; and (c) improve public awareness and understanding of various stakeholders about climate change, vulnerabilities, and impacts in order to increase stakeholder willingness to take actions	National Strategy on Climate Change (2010) and INDC (2015)
Malaysia	Policy Statement of Malaysia reveals its desire 'to ensure climate resilient development to fulfil National aspirations for sustainability'. Objectives of the country are to (a) mainstream climate change for strengthening competitiveness and improvement of quality of life; (b) integration of climate change responses into policies, plans, and programmes; and (c) strengthening of institutional and implementation capacity	National Policy on Climate Change (2009)

(continued)

Table 8.1 (continued)

Country	Adaptation vision/goal/objective	Source
Myanmar	Vision of Myanmar is to be 'a climate-resilient, low-carbon society that is sustainable, prosperous, and inclusive, for the well-being of present and future generations'	National Climate Change Strategy and Action Plan (2017)
Philippines	Its vision is to be 'a climate risk-resilient Philippines with healthy, safe, prosperous, and self-reliant communities, and thriving and productive ecosystems'. Its goal is to build the adaptive capacity of communities and increase the resilience of natural ecosystems to climate change	National Framework Strategy on Climate Change (2010–2022)
Singapore	Vision of Singapore is to be 'a climate resilient city that is well positioned for green growth'. It aims to take steps collectively as a nation today that will ensure Singapore remains a vibrant and liveable nation, a home that is flourishing and enduring for this and future generations	National CC Strategy (2012)
Thailand	Vision of the Royal Thai Government is to achieve climate resilience and low carbon growth in accordance with sustainable development agenda. It sets its mission to build climate resilience for Thailand's development by mainstreaming climate change adaptation into development planning of all sectors and levels	Climate Change Master Plan (2013–2050)
Vietnam	Strategic objectives of the Government of Vietnam are (a) to assess climate change impacts on sectors and regions in specific periods; (b) to develop feasible action plans to effectively respond to climate change in the short term and long term to ensure sustainable development of Vietnam; (c) to take opportunities to develop towards a low-carbon economy; and (d) to join the international community's efforts in mitigating climate change and protecting the climatic system	National Target Programme to Respond to Climate Change (2008)

Note: Brunei and Timor Leste have yet to develop their adaptation vision/goal

landlocked country). These priorities are consistent with the countries' adaptation goals/targets which stress the importance of sustainably managing natural resources, leading to economic development. Agriculture, and water resources are highly significant for Least Developed Countries (LDCs) in the region, i.e. Cambodia, Lao PDR, Myanmar, and Timor Leste, because more than half of their labour force is still reliant on agriculture (CIA 2015).

Nearly all SEA countries also recognise the need for resilient energy systems, although this sector is often linked with climate change mitigation. This demonstrates a high level of awareness of the importance of energy security especially in emerging economies. SEA countries have also highlighted several cross-sectoral issues including capacity building, mainstreaming, early warning systems, and awareness raising. Forestry and biodiversity are also regarded with high importance by more than half of the countries. For a region endowed with rich biodiversity and undergoing rapid economic growth, SEA countries prove that they are aware of and value the role of biodiversity and forestry to attain sustainable development.

Table 8.2 CCA Priority Sectors in SEA

Country	Agriculture	Water resources	Coastal/marine resources	Forestry	Biodiversity	Health	Energy	Infrastructure	Industry	Transport
Brunei Darussalam	✓	✓	✓	✓	✓	✓				
Cambodia	✓	✓	✓		✓	✓				
Indonesia	✓	✓	✓	✓		✓	✓	✓		
Lao PDR	✓	✓		✓		✓	✓		✓	✓
Malaysia	✓	✓	✓	✓	✓	✓	✓			
Myanmar	✓	✓	✓	✓	✓	✓	✓		✓	
Philippines	✓	✓	✓	✓	✓	✓	✓		✓	
Singapore		✓	✓		✓	✓	✓	✓		
Thailand	✓	✓	✓	✓	✓	✓				
Timor Leste	✓	✓	✓	✓		✓	✓	✓		
Vietnam	✓	✓	✓	✓		✓	✓			✓

Source: Revised table from APAN- ICLEI (2015)

8.3 Adaptation Trends in Southeast Asia

It is well recognised that certain CCA actions/initiatives are well aligned with development aspirations and sustainable resource management. The World Resources Institute (2009) presents a continuum of adaptation options, from those focused on addressing vulnerabilities to those focused on impacts (Fig. 8.2). The continuum is categorised into four main parts:

1. Addressing the drivers of vulnerability—objectives overlap with normal development activities because it is mainly targeted at improving human conditions (e.g. poverty reduction, education).
2. Building response capacity—capacity building efforts that would be the basis of targeted actions (e.g. improved planning and communications processes, early warning systems).
3. Managing climate risk—inclusion of climate information into decision-making (e.g. collection of local climate data, climate risk assessments).
4. Confronting climate change—specifically targeted at addressing climate change impacts (e.g. crop diversification, reducing the risk of glacial melting).

An examination of CCA strategies and actions that were identified for each priority sector in the country reports, NAPAs, and National Communications would reveal CCA focus and trend in the region. Most existing and proposed adaptation initiatives in the region are categorised in the middle of the continuum, with 'building response capacity' being more dominant. Example initiatives under this category include vegetation planting for flood and windstorm protection, strengthening disaster preparedness and response capacity of community, and establishment of conservation corridors. This is consistent with Malaysia's thrust stating that priority is given to adaptation based on wise resource management (Malaysia INDC).

For LDCs, however, several adaptation initiatives fall under 'addressing drivers of vulnerability' part. Examples of such initiatives include safe water supply for rural communities, promotion of household integrated farming; development and improvement of community irrigation systems, and development of healthcare

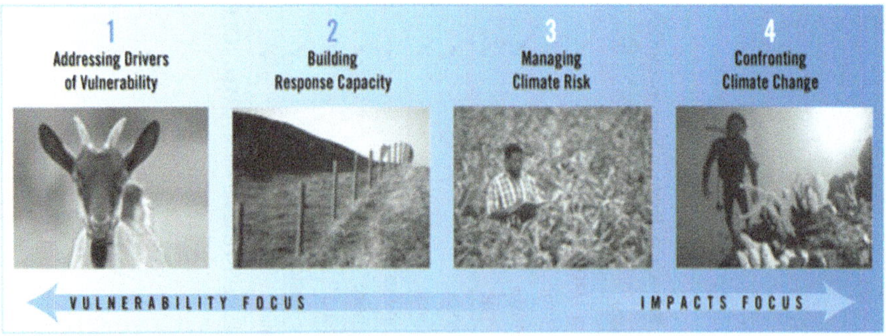

Fig. 8.2 Adaptation Continuum. Source: WRI (2009)

centres. To clarify, initiatives that directly 'confront climate change' are already being implemented in the region, such as the use of drought/saline water—resilient rice varieties and coral bleaching mitigation efforts. In general, as many SEA countries are still developing, there is preference of no- and low-regrets strategies or those initiatives with co-benefits, often overlapping with development priorities.

8.3.1 Policy, Strategy, Legal Provision, and Governance

Most SEA countries have progressed in formulating climate change strategies and action plans at the national level. Table 8.3 presents an overview of past and existing climate change strategies and action plans of each SEA country. For several countries, CCA strategies are included in the overall CC strategy/action plan (combined with mitigation). Indonesia, Philippines, and Vietnam (for a single sector) have created separate CCA strategy and/or action plans. Also, the LDCs in the region have all submitted their NAPAs, which is a key step in advancing adaptation efforts as it serves as a key to access CCA financing. Among the countries, Brunei seems to be lagging in terms of adaptation planning.

In terms of climate change policy, to date, only the Philippines (Climate Change Act of 2009) enacted a national law on climate change. Malaysia has its National Policy on Climate Change, which serves as a blueprint of the country's climate change actions. Cambodia announced that it will develop a climate change legal framework. Similarly, Myanmar intends to create a National Climate Change Policy in 2016 (INDC).

8.3.2 Institutional Arrangement

Formal institutions that deal with climate change have been established in each country in the region (Table 8.4). These institutions either exist as independent bodies or special divisions within the Ministry/Department of Environment. Whilst recognising the cross-cutting nature of CCA, many of these institutions are composed of representatives from various ministries. To date, only Timor Leste does not have a formal climate change unit or secretariat. In the Philippines and Singapore, the climate change secretariat is a separate body led by the Head of State. The common roles of climate change bodies in the region include: (1) to lead in mainstreaming climate change in government plans and programmes, (2) to develop strategies and guidelines for the achievement of climate change goals, and (3) to coordinate actions among related sectors (both public and private) for a unified action (APAN-ICLEI 2015).

Table 8.3 Climate change policies, plans, and strategies of SEA countries

Country	Climate change plan/strategy
Brunei Darussalam	• No climate change strategy and plan (to date). However, sectoral adaptation plans exist, but a national integrated and coordinated adaptation plan has yet to be developed (INDC)
Cambodia	• National Adaptation Programme of Action (NAPA) 2007 • Sectoral Climate Change Action Plan (2014–2018) • Cambodia Climate Change Strategic Plan (2014–2023) • National Strategic Plan on CCA (to be developed)
Indonesia	• National Action Plan on Climate Change (2010) • Climate Change Sectoral Roadmap (2010) • National CCA Action Plan (RAN-API) (2013)
Lao PDR	• National Strategy on Climate Change (2010) • National Adaptation Programme of Action (2009)
Malaysia	• National Policy on Climate Change (2010). A national adaptation plan would be developed (INDC 2015)
Myanmar	• National Adaptation Programme of Action (NAPA) 2013 • National Climate Change Strategy and Sector Action Plans 2016 • National Adaptation Plan will start in 2016 (INDC 2015) • National Climate Change Policy is expected to be approved in 2016
Philippines	• Climate Change Act of 2009 • National Climate Change Action Plan (2011–2028) • National Framework Strategy on Climate Change (2010) • Philippine Strategy on Climate Change Adaptation (2010–2022)
Singapore	• National Climate Change Strategy (2012)
Thailand	• National Strategy on Climate Change (2008–2012) • Thailand Climate Change Master Plan (2013–2050) • Thailand Adaptation Plan (2014–2018) (under development)
Timor Leste	• National Adaptation Programme of Action (NAPA) (2011) • Development of CCA Action Plan is urgently needed (INC)
Vietnam	• National Climate Change Strategy (2011) • National Target Program to Respond to Climate Change (NTP-RCC) (2013–2015)/NTP-RCC (2008–2012) • National Action Program on CC (2012–2020) • Action Plan Framework for Adaptation and Mitigation of CC in the Agriculture and Rural Development Sector Period (2008–2020)

*Note: Brunei has yet to conduct its vulnerability/impact assessment (Yusra 2013). It has not formulated its climate change policy, plan, or strategy (Brunei HFA)

8.3.3 Integration and Mainstreaming

Several SEA countries have incorporated climate change in their national development plans. The status of integration of climate change in each country is presented in Table 8.5. Cambodia's National Strategic Development Plan has included climate change in more than 10 sectors including rural development, research and development in agriculture, health, gender mainstreaming, and disaster management. The

Table 8.4 Climate change offices of SEA countries

Country	Climate Change Institutions/Bodies
Brunei Darussalam	• National Council on Climate Change headed by the Ministry of Development and the Energy Division at the Prime Minister's Office
Cambodia	• National Climate Change Committee. Climate Change Department within Ministry of Environment provides secretarial function and Climate Change Technical Teams provides technical advice and support
Indonesia	• Director-General of Climate Change under the Ministry of Environment and Forestry
Lao PDR	• National Steering Committee on Climate Change, Technical Working Groups, and National Climate Change Office are main bodies involved in climate change decision-making process
Malaysia	• Environmental Management and Climate Change Division (within the Ministry of Natural Resources and Environment) • Green Technology and Climate Change Council (headed by the President) • National Steering Committee on Climate Change • Working Committee on Adaptation is anchored by the NRE
Myanmar	• Climate Change Division under Environment Conservation Department of Ministry of Natural Resources and Environment Conservation (MoNREC)
Philippines	• Climate Change Commission (Office of the President) • Climate Change Office in the Department of Environment and Natural Resources
Singapore	• National Climate Change Secretariat • Inter-Ministerial Committee on Climate Change (for international negotiations) • Resilience Working Group • Centre for Climate Research Singapore
Thailand	• National Committee on Climate Change (chaired by the Prime Minister)
Timor Leste	• National Adaptation Programme of Action (NAPA) Team
Vietnam	• National Steering Committee for the implementation of UNFCCC and Kyoto Protocol

Royal Government of Cambodia (RGS) also promoting mainstreaming of CCA in sub-national plans and finance systems. In Myanmar, several ministries are taking steps for mainstreaming CCA in their planning processes (INDC 2015). In the Philippine Development Plan, resilience to climate change was highlighted in three target sectors—competitive and sustainable agriculture and fisheries sector, accelerating infrastructure development, and conservation, protection, and rehabilitation of environment and natural resources. The Philippine government also aims for holistic resilience building by ensuring the close integration of CCA and DRR in plans and programmes—spanning several sectors and at different levels of government. In Singapore, building resilience to climate change is identified as an element to achieve 'a vibrant and sustainable city'. All these demonstrate that most SEA countries regard climate change as crucial in attaining sustainable development. It is also interesting to note that the countries (e.g. Indonesia, Lao PDR, and Cambodia) are aware of CCA's role in achieving improved natural resources management and increased food, water, and energy resilience (Indonesia and Lao PDR). Since many

Table 8.5 Climate change in National Plans and Strategies

Country	Development plan	Status of integration
Brunei Darussalam	10th National Development Plan Brunei Vision (2035)	No mention of climate change
Cambodia	National Strategic Development Plan (2014–2018) National Strategic Plan on Green Development (2013–2030)	Mainstreamed climate change
Indonesia	National Medium-Term Development Plan (2010–2014) National Development Planning: Response to Climate Change (Yellow Book)	Integrated climate change mitigation and adaptation
Lao PDR	National Development Socio Economic Development Plan (2010–2015) Seventh National Social Economic Development Plans National Capacity Self-Assessment (NCSA)	Mainstreamed climate change
Malaysia	10th Malaysia Plan 2011–2015: Chapter: Valuing the Nation's Environmental Endowments National Physical Plan 2	Mentioned adaptation strategy
Myanmar	National Sustainable Development Strategy for Myanmar (2009) Myanmar Climate Change Policy (2017)	Identified six priority sectors for climate change adaptation
Philippines	Philippine Development Plan (2011–2016): Mid-term Philippine Development Plan	60% of the chapters are climate-proofed; actions required are consistent with NCCAP
Singapore	Sustainable Singapore Blueprint (2015)	Mainstreamed climate change
Thailand	11th National Economic and Social Development Plan (2012–2016)	Upgrade the ability to adapt to climate change
Timor Leste	Timor Leste Strategic Development Plan (2011–2030)	Climate change within Environment under the Social Capital Section
Vietnam	The National Strategy on Environment Protection to 2020, with Visions to 2030	General objectives—to raise the capability of responding climate change

SEA countries are experiencing rapid economic growth, recognition of climate change impacts on development is imperative.

8.3.4 *Finance*

High adaptation costs pose a serious challenge to developing SEA countries, which must allocate scarce resources to competing development agendas. Fortunately, there are several financing sources which aid countries to reduce inherent vulnerabilities and increase both resilience and adaptive capacities to the impacts of climate

Country	USD (in millions)
Cambodia	130.7
Indonesia	13.7
Lao PDR	25.1
Myanmar	19.3
Philippines	14.1
Thailand	0.9
Timor Leste	22.0
Vietnam	20.1

Fig. 8.3 Multilateral CCA recipients in SEA (Climate Funds Update 2016)

change. A significant amount of CCA financing in the region comes from bilateral and multilateral sources. Major financing sources in SEA include the Pilot Program for Climate Resilience (PPCR), Least Developed Countries Fund, and Adaptation Fund (Climate Funds Update 2015), and some are under the bilateral Official Development Assistance. Figure 8.3 shows the distribution of CCA financing (multilateral sources) in the region. It can be observed that Cambodia receives bulk of the funds. Also, as expected, Brunei, Singapore, and Malaysia are not included because of their economic status. Developing countries in the region, especially the LDCs, are still dependent on financial support from various funding agencies. For example, in Myanmar, although CCA is included in all sectors of the national budget, the government states that 'the scale of the adaptation effort, considering the current level of capacities, and the trend of worsening changes in climate, means major support and investment is required' (INDC 2015). For countries like Indonesia and Cambodia, which receive a significant amount of climate financing, establishing climate change trust funds became imperative for accountability and better management/mobilisation of resources. Cambodia reports that it has raised around US$ 250 million from multi/bilateral sources to implement projects addressing climate change. The PPCR has assisted the Royal Government of Cambodia to strengthen three priority sectors—water, agriculture, infrastructure (Cambodia NSDP p. 36).

Some countries in the region are willing to co-finance adaptation initiatives and have allocated a portion of the national budget for CCA initiatives. For example, the Philippines has its People's Survival Fund, a pool of money which can be accessed by local governments to implement resilience building initiatives. It is also implementing a climate expenditure tagging system that incorporates climate change initiatives into the annual budgeting process (INDC 2015). Another example is Singapore, which invests in climate change research.

8.3.5 Implementation Capacity

Several SEA countries receive multilateral CCA financing and have experiences in implementing CCA projects. This can be regarded as progressive in implementing CCA initiatives and to a certain extent capacity do exist to implement CCA projects/ programmes.

In terms of assessing the effectiveness of adaptation initiatives, most countries have yet to establish a monitoring and evaluating (M&E) framework although countries recognise its importance in their climate change strategies/action plans. As a step forward, Cambodia and the Philippines have set up specific indicators in their respective development plans in order to measure effectiveness of the implementation of planned adaptation actions. The Philippines, through the support of GIZ, was able to come up with its M&E framework and has identified specific indicators for each identified priority sector in the NCCAP. Similarly, Vietnam released its key M&E indicators in terms of planning, poverty alleviation, storm shelters and communication equipment, mangrove cover, access to clean water and healthcare (INDC 2015). On the other hand, Myanmar states that it will develop an appropriate mechanism for monitoring of climate vulnerability, funds allocated for adaptation and the results of adaptation actions (INDC 2015).

8.3.6 Technology

Several countries (e.g. Vietnam, Thailand, and Philippines) have adopted farming technologies such as the use of crop varieties that are resistant to drought and saline water. Being in the sub-region, the International Rice Research Institute (IRRI) has a prominent role in aiding SEA countries to promote and conduct more research on climate resilient agricultural technologies.

In the Philippines, there has been a big investment in reliable weather forecasting systems to predict the intensity and speed of typhoons that would enter the Philippine territory, which is the basis for the government to issue early warnings for the people. It is also common for SEA countries to use the Geographic Information System (GIS) to map out vulnerabilities to climate change impacts such as flood, drought, and sea level rise. This is a valuable tool to improve and better inform planning and policy-making. In Singapore, research and technology development are being harnessed to improve and augment water supply (i.e. NEWater, desalination).

Thus, adoption and use of technologies exist. Moreover, countries have established institutions that would advance science and technologies, such as the Department of Science and Technology (the Philippines), the Thailand Institute of Scientific and Technological Research (Thailand), and the State Ministry of Research and Technology (Indonesia).

8.3.7 Decision-Making Tools and Institutionalisation

All SEA countries have conducted at least a preliminary nationwide vulnerability and risk assessments of priority sectors. It is evident that the LDCs highly depend on external support for conducting vulnerability and impacts assessments due to lack of resources and technical capacities. For example, the National Technical Committee of the UNDP/GEF sponsored Cambodia's Enabling Activity Project to conduct assessment of potential climate change impacts on priority sectors (Cambodia INC). In Lao PDR, sensitivity and exposure index and adaptive capacity index of villages have been developed and identified, coupled with probability of climate hazards (Lao PDR SNC). On the other hand, Brunei, which is still in the early stages of implementing CCA, have yet to conduct a comprehensive vulnerability and impact assessments, which impedes formulation of CCA plans/strategies (Yusra 2013; Brunei INDC).

For countries such as Indonesia, Philippines, and Thailand, studies related to climate change impacts and vulnerabilities conducted by research institutes, universities, and non-governmental organisations are valuable in aiding national governments in the decision-making processes. On the other hand, having the resources, Singapore established the Centre for Climate Research Singapore (CCRS) to enhance knowledge about climate science and modelling. Similarly, Malaysia has developed its own Regional Hydro-Climate Model for Peninsular Malaysia (RegHCM-PM), which is an attempt to downscale climate projection from a global model.

8.3.8 Awareness, Knowledge, and Access to Information

Climate change issues cut across several sectors and its impacts are experienced at the local level. Therefore, for CCA to be effective it is imperative to engage multi-stakeholders. In recognition of this, several SEA countries have launched public outreach and education programmes focused on climate change. In Cambodia, the Ministry of Education, Youth, and Sports has started to integrate CC issues into the formal and non-formal education systems (Cambodia NSDP p. 36). The Philippines is active in engaging and empowering the youth on climate change through various campaigns such as the #NowPH: One Million Filipino Youth Voices for Climate Action Campaign. In Myanmar, the Monsoon Forum is held annually to enhance climate forecasting. Myanmar Climate Change Alliance (MCCA) project works with media to increase awareness of CSOs and other stakeholders by increasing broadcasts and news on climate change and production of awareness materials throughout 2016–17 (INDC 2015). Singapore has incorporated climate change in subjects such as 'GCE "A" Levels General Paper, Economics, Geography, and the sciences' (TNC 2015). These are just a few of the many ongoing programmes and projects in the region. In addition to government projects and programme, different non-government organisations and civil societies have also aided in the production of educational materials targeted at several different audiences.

In terms of disseminating weather data and warnings, almost all countries (except Timor Leste) employ online platforms through the websites of their respective weather bureaus. Governments of the Philippines, Thailand, Malaysia, Indonesia, and Myanmar also reach out to the public through social media and mobile applications such as Facebook, Twitter, Widget, Podcast, Android App, Google Play, and Instagram.

In an interesting case of updating an early warning system, the Philippine Atmospheric, Geophysical and Astronomical Services Administration (PAGASA) needed to consider the increasing intensity of typhoons in the past decade. Just recently, the weather bureau has developed a new set of storm signals, giving way to severe tropical storm (storm signal #3) and super typhoon (storm signal #5). This was in response to speculations from local authorities saying that the warning ('typhoon') given for Typhoon Haiyan in 2013 was not strong enough to give an impression of great urgency and emergency.

8.4 Gaps and Areas of Adjustment

8.4.1 Policy, Strategy, Legal Provision, and Governance

Policies have been put into place, and strategies and concrete action plans have been formulated in many countries at the national level. However, not much reporting is being done about how CCA is advancing in terms of policy/legal framework at the sub-national or local levels. Countries have yet to give more attention to the vital role of local governments which are at the frontline, directly experiencing climate change impacts. Challenges also revealed uncoordinated planning and the tendency to use a short-term planning horizon, lack of on-the-ground adaptation interventions, and weak or absence of M&E mechanisms that would assess effectiveness of implementation of policies/plans/strategies. Further, Brunei and Timor Leste, indicated the prevailing policy/regulatory constraints such as lack of regulatory framework (no specific climate change policies/plans/programmes), limited integration of climate change in national policies and programmes, and poor integration of vulnerability and adaptation strategies into various sectors. Brunei, in its INDC, recognises the need to 'establish and promote the appropriate legislative framework' and to develop a 'national integrated and coordinated national plan'. In Myanmar, sectors prioritised in NAPA requires further elaboration under the framework of an overall adaptation plan, including the costing and monitoring framework that highlights the roadmap to resilience (INDC 2015).

8.4.2 Institutional Arrangement

Coordinating several ministries and agencies at various levels is not an easy task. At times, lack of coordination can lead to fragmented initiatives, turf issues, and overlapping responsibilities. For example, in the Philippines, climate change initiatives tend to be dispersed across several bodies working on climate change (refer to Table 8.4). This reduces efficiency and effectiveness of action and resource use. Indonesia acknowledges that it needs to strengthen its institutional capacity (INDC 2015). This is particularly important after the reorganisation of climate change-related offices (e.g. abolishment of DNPI) at the start of the Jokowi administration. To attain the adaptation goals/targets, it is necessary for countries to enhance coordination not only among ministries and agencies but also include provincial and local governments, as well as non-government actors. It is important to note that there is very little mention of the private sector engagement in adaptation efforts. The Philippines mentioned the need for an enabling environment for private sector engagement in CCA (INDC 2015). Businesses present a high potential to augment climate adaptation financing and engage in risk transfer mechanisms. This is something SEA countries may take advantage of. Multi-stakeholder cooperation and collaboration through vertical and horizontal integration is key to having a holistic approach to CCA in the sub-region.

8.4.3 Integration and Mainstreaming

Integration of CC into policies, plans, and programmes is necessary to achieve the adaptation vision. Integration, however, does not mean the mere mention of climate change in official documents. In analysing plans and policies, there are cases that show that even though CCA is mentioned, bulk of the text is related to disaster risk management, natural resource management, or poverty reduction. Thus, there is a need to highlight CCA characteristics, particularly the need to understand climate vulnerabilities based on future climate projections.

While SEA countries are aware of the significance and most have advanced in mainstreaming CCA in policies and programmes, shared challenges that persist include vague integration of CCA into policies and programmes (not clear on who will implement and how), uncoordinated sectoral and national-local plans, skewed approach wherein CCA policies/programmes focus mainly on post-disaster emergency relief, and fragmented approach to poverty reduction and CCA, among others. This is particularly difficult for countries which have just started implementing CCA, such as Brunei and Myanmar. The two countries are recognising the need to mainstream and integrate CCA in development plans. Brunei states in its INDC that it needs 'to strengthen CCA effort by promoting the mainstreaming of sectoral adaptation plans into a national holistic and coordinated plan', while Myanmar declares it is 'critically urgent that CC concerns are integrated into all future national and sectoral development plans and programs'. In addition, in many cases, mainstreaming and integration remains at the national level. Although there is recognition that it is equally important to trickle the process down to the local level, lack of understanding and awareness about the issues, and limited guidance on how to mainstream CCA, how to institutionalise processes and delegate responsibilities, and how to create concrete local CCA plans, remain prominent.

8.4.4 Finance

Plans and policies should be backed by financial resources to move toward implementation. Bulk of climate adaptation financing in the region comes from public sources. LDCs highly rely on international CCA financing due to limited government funds that must compete with other development priorities. However, there are challenges in accessing international climate financing such as cumbersome processes, numerous requirements, and proposal writing skills. Eligibility requirements, particularly in terms of economic status, prevent some countries to apply for assistance such as Malaysia and Brunei, wherein there is an enormous potential to further boost CCA. These countries may need a new mechanism to access international support.

In Fig. 8.3, Cambodia has been able to access a significant amount of CCA financing, whilst other LDCs are lagging. This implies that Cambodia has a comparative advantage in accessing CCA financing (e.g. institutions in place). This highlights the need to further capacitate other LDCs in the region. It is important that countries can show competence not only in implementing CCA but also in managing CCA financing.

It is highly likely that public financing would not be enough to meet future demands, thus, countries would need to look for alternative adaptation financing sources such as allocation of internal funds, increased engagement of the private sector, adoption of payments for ecosystem services (PES). Also, for any adaptation project, financial accountability and value for money are important to make sure that limited financial resources are allocated properly.

Moreover, it is important to note that most international CCA financing sources are coursed through national governments. This is very crucial because the ones which implement laws and plans on the ground are often the local governments. It is important to note this gap and ensure that local governments have both the knowledge to implement CCA initiatives and the skills to effectively manage finances.

8.4.5 Implementation Capacity

Lack of technical expertise and knowledge is a big challenge for SEA countries. As a clear example, all LDCs were supported by UN agencies (together with GEF) to come up with their NAPAs. Lao PDR mentions it was 'heavily dependent on external experts' (SNC). Myanmar and Timor Leste, both who have just recently submitted their INCs, admittedly declared the lack of capacity and the need for external support for capacity building. Other countries were also supported to produce their National Communications, emphasising the lack of technical expertise and capacity in these countries. This gap persists up to the present. In fact, almost all countries (except Malaysia and Singapore), which have submitted their INDCs, specifically stated the need for more capacity building in CCA. Common implementation gaps in the region include lack of institutional and administrative capacity, especially at the local/community level, absence of comprehensive monitoring of actions, lack of technical skills/expertise in climate science, modelling, and socio-economic impact assessment, low understanding of uncertainties and limitations of projections/scenarios, shortage of technical experts, and little knowledge of risk management measures (risk sharing/transfer mechanisms).

8.4.6 Technology

A summary of the Technology Needs Assessment reports (TNAs), submitted to the UNFCCC in 2012 by some Southeast Asian countries, is presented in Table 8.6. Water and agriculture remain the priority sectors for technological needs. Interestingly, Myanmar and Timor Leste reiterate technological needs for the same sectors in their NAPAs.

SEA countries need to give attention and address the following in order to advance in achieving their adaptation vision: difficulty in downscaling available data (global/regional to local scenarios), lack of scientific and technical data (climate data) or incomplete database (especially locally relevant and useful information), lack of methodologies, development models and tools for technological needs (i.e. impacts and vulnerability assessment, and appraisal of adaptation measures), lack of means to effectively translate climate projections into sectoral impacts, and high uncertainties and limitations of climate scenarios/simulation.

8.4.7 Decision-Making Tools and Institutionalisation

There is still a big gap for several SEA countries in terms of limited or incomplete climate data, lack of downscaled climate projections, uncertainties in climate simulation, lack of locally relevant methodologies and tools, and lack of expertise to process data acquired. All these are valuable in providing useful information for well-informed adaptation planning.

Thailand states that it 'has no official and effective means of data collection, data transfer, and database management at regional level' (TNA 2012). Indonesia emphasises the need for 'enhanced actions to study and map regional vulnerabilities as the basis of adaptation information systems' and 'certainty in spatial planning and land

Table 8.6 Summary of technology needs in selected SEA countries

Country	Priority sector	Priority technologies
Cambodia	Water Household safe water supply Community water supply	• Rainwater harvesting from rooftops • Wells for domestic water supply • Small reservoirs, small dams, and micro-catchments
	Coastal resources	• Mangrove management
Lao PDR	Water	• Early warning system • Disaster impact reduction fund • Climate change-oriented irrigation • Water supply system
	Agriculture	• Livestock disease prevention and control • Agricultural development subsidy mechanism • Climate resilient rural infrastructure • Crop diversification
Thailand	Agriculture	• Climate forecast and early warning systems • Crop improvement • Precision farming practices • Post-harvest technology • Animal nutrition and feed technology
	Water resources	• Environmental observation • Weather and hydrological modelling • Flood and drought risk management • Operation of water infrastructures • Community water resource management • Integrated urban water resource management • Early warning systems
	Modelling (cross-sectoral)	• National data centre • National data transfer/management process • Weather Research and Forecasting (Advanced Research WRF)
Vietnam	Agriculture	• Plant genetics/Plant breeding • Shifting from rice to upland grains • Shifting from triple cropping to double cropping and a shrimp/fish/poultry crop
	Forestry	• Plant science/plant genetics • Agroforestry
	Coastal resources	• Integrated coastal zone management • Sea dykes • Coastal wetland rehabilitation • Flood warning system
	Water	• Harvesting rooftop rainfall for household usage • Harvesting runoff water • Irrigated river basin management

use, including land tenure' (INDC 2015). On the other hand, Singapore highlights in its INDC: 'Singapore's location within Southeast Asia, where there is a lack of observational climate data, is also a challenge to local climate modelling work in projecting climate change impacts. These characteristics make adaptation planning in Singapore both complex and challenging'. All these imply the need for improved regional cooperation and knowledge sharing, particularly in this aspect.

SEA countries are advised to invest in acquiring relevant and up-to-date information, conduct periodic vulnerability, risk and economic assessments, and take advantage of existing tools which can be applied in the country context. International research institutions and universities have an enormous potential in promoting climate modelling and science-based vulnerability and impact assessments in the SEA. Moreover, countries need to move beyond projects by putting a system in place to review processes, Vas, and address emerging issues.

8.4.8 Awareness, Knowledge, and Access to Information

For the region to attain its adaptation goals, countries need to further bolster public awareness and knowledge on climate change. In their INDCs, Myanmar and Brunei state the importance of engaging multiple stakeholders across sectors and at all levels through improved communication and education. Similarly, Indonesia emphasise the value of sharing best practices by 'scaling up diversity of traditional wisdom and innovative climate change efforts (government, private sector, and communities)' (INDC).

Awareness raising and education are especially essential for rural communities, which are among the most vulnerable to climate change as they dependent on natural resources that are vulnerable to climate change impacts. The SEA countries need to take advantage of existing educational materials and videos which are available to use for free. At the government level, particularly in local governments, all employees (not only in the environment or planning division) should be included in information and education programmes so that they will be able to integrate climate change in their operations.

Table 8.7 provides an overall summary of status of climate change adaptation against eight indicators and future needs for enhancing adaptation actions at several levels.

8.5 Readiness for Enhanced Adaptation Actions

Salamanca and Nguyen (2016) wrote that 'the concept of "adaptation readiness" provides a framework to describe where countries are with regard to formulating and achieving their goals of adapting to the impacts of climate change and

Table 8.7 Summary of status and future need for advancing climate change adaptation

Indicators	Achievements to date	Moving forward
Policy, strategy, legal provision, and governance	Formulation of adaptation strategies and action plans (except Brunei)	Linked/coordinated policies, plans, and programmes (umbrella framework, holistic approach) Long-term visioning Develop monitoring and evaluation mechanisms Improve local climate governance Constant review and updating of strategies and plans Mainstream CCA and DRR
Institutional arrangement	Established CC bodies/offices. Bodies have been established	Better coordination, collaboration, and communication among relevant ministries/offices (horizontal) and at all levels (vertical) Clear roles/responsibilities Enhanced multi-stakeholder partnerships and regional cooperation (e.g. ASEAN) Strengthen institutional frameworks/ mechanisms
Integration and mainstreaming	CC mainstreamed into countries' development and sectoral plans Recognise the importance of CCA and resilience building	Increase knowledge and capacity on CCA and DRR towards a holistic approach to resilience building Holistic approach to CCA through systematic programming
Finance	Existing efforts to mobilise internal funds for CCA initiatives Improving access to CCA financing	Improve climate finance management Promote transparency and accountability Mobilise private sector financing and advocate for PES Enhance climate financing at the local level Maintaining special financial support for LDCs
Implementation Capacity	CCA projects and programmes in countries Increased capacity and knowledge on CCA through technical assistance, projects, and trainings	Enhance technical expertise and knowledge Build capacities of local authorities and communities Trickle down opportunities at sub-national levels Move beyond project-based approach Develop research and capacity on CC Effective implementation of CCA strategies and plans

(continued)

Table 8.7 (continued)

Indicators	Achievements to date	Moving forward
Technology	Adoption and use of various technologies to better deal with CC risks and impacts	Improve access to low cost/ affordable, yet easy to understand technologies; real-time monitoring Invest in locally appropriate technologies Increase international cooperation for technology transfer
Decision-making tools and institutionalisation	Conduct of vulnerability and impacts assessments for key sectors—in many cases with external support	Include economic assessment (e.g. CBA) of adaptation options Conduct information-generating activities (e.g. research, community-based VA) Methods for understanding social impacts Guides for climate risk management Triangulation of research and assessment results Modelling for several sectors and intersectoral VA
Awareness, knowledge, and access to information	Launching public outreach (education, information, and communication) programmes Mainstreamed into school curriculum	Enhance multi-stakeholder engagement Better communicate science to policy-makers and local communities Increase access to and the ability to interpret information Comprehensive local communications strategy Enhance and promote research outreach

implementing key policy priorities'. Using the eight parameters presented in detail in the previous section, SEA countries are at varying levels of adaptation readiness.

The SEA countries recognise the need to adapt to the adverse impacts of climate change. This is evident in each country's climate change action plans, which are often approved by the highest national-level authority. Some plans are short- to medium-term while others account for longer time periods such as the Philippines' National Climate Change Action Plan covering 2011 to 2028, and Thailand's Climate Change Master Plan spanning from 2013 to 2050. However, translating these plans and strategies into tangible programmes and projects is altogether a different concern.

Existing institutional arrangements show that most SEA countries classify climate change as a function under the environment ministry. Some countries such as Singapore and the Philippines have established separate entity to oversee and coordinate climate change concerns. In January 2015, Indonesia's National Council on Climate Change was merged by President Jokowi through a Presidential Decree with the Environment and Forestry Ministry. Regardless of whether the climate change function is subsumed under the environment ministry or a separate entity oversees climate change-related matters, the key to enhance adaptation action is through clear and coordinated multi-sectoral involvement and participation.

Review of national reports including INDCs revealed that SEA countries acknowledge the importance of linking climate change action plans with other development and sectoral plans. The Cambodia Climate Change Strategic Plan (CCCSP) 2014–2023 noted that 'the action plans… will not be developed as stand-alone "silos" within the respective ministries'. Myanmar's INC stated that integration of climate change into development plans and programmes are considered vital in sustaining the country's 'economic development in harmony with social and environmental integrity'.

However, both countries indicated problems associated with integration and mainstreaming. In the CCSP, it was mentioned that clear procedures to integrate climate change in development plans are still lacking. In Myanmar, the designated body to oversee integration does not have the authority and capacity to integrate environment and development across different levels (regional, township, villages) and among different ministries and agencies.

The International Fund for Agricultural Development—IFAD (2014) wrote that in Vietnam, the NTP-RCC, approved in 2008, serves as the country's umbrella programme and framework implemented by the Ministry of Environment and Natural Resources (MoNRE). The MoNRE has already developed monitoring indicators and policy matrix for implementing the NTP-RCC. However, possible challenges 'seem to lie more with the various line ministries' tasked to develop sectoral strategies and action plans. The country's INDC further reiterated that 'the legal framework for integrating climate change issues into national Socio-Economic Development Plans is still limited. There is still ineffective coordination between line ministries, sectors, and localities to address multi-sectoral and interregional issues'.

Climate financing is another dimension that needs examination in relation to enhancing adaptation actions. Based on the 2013 report of the OECD-DAC as cited by the Adaptation Committee,[2] Asia and Africa combined to get about two-thirds of adaptation finance (Fig. 8.4). At the country level, three out of the top ten countries receiving the most adaptation-related aid are located in SEA, i.e. Vietnam, the Philippines, and Indonesia. While there are efforts to mobilise funds internally, ICLEI SEAS (2013) noted that most adaptation financing in the region comes from bilateral and multilateral sources, with heavy reliance on grants in the form of official development assistance.

Except for Singapore and Brunei, all countries in SEA are classified as developing. This indicates that there are existing gaps in terms of implementation capacity, availability of decision-making tools, and technologies. Where these are available,

[2] Seventh meeting of the Adaptation Committee: Bonn, Germany, 24–27 February 2015. *Synthesis paper on the means of implementation.*

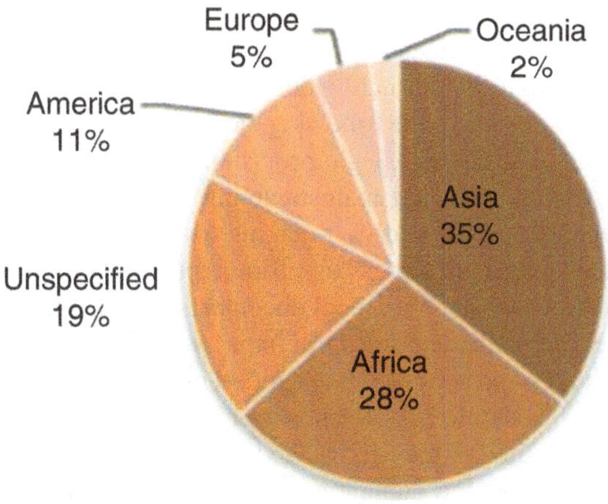

Fig. 8.4 Geographic allocation of adaptation finance (OECD-DAC CRS, 2010–12)

they are often insufficient and tend to be skewed toward highly urbanised cities as well as regional centres of development or remain at the national level.

Thailand's Second National Communication (SNC) revealed that adaptation research is still in its infancy stage. Main problems on adaptation are linked with (1) research techniques to prioritise key sectors and analyse best alternatives for adaptation; (2) integration of adaptation options into the socio-economic development of risk-prone communities; and (3) technology options for villages in disaster-prone areas. On the other hand, Lao PDR mentioned the lack of long-term historical data and appropriate climate scenarios for impact analysis are major constraints.

In the Philippines, adaptation concerns relate more to data collection and level of granularity. These include need for higher resolution scenarios, at least up to the provincial level; unavailability of daily projected values for impact assessments; shortage and inaccessibility of non-climatic data (socio-economic, environment, and other related information); and outdated base maps (topographic, land use, and resource base, etc.).

Singapore can be a model for its neighbouring countries in terms of bridging capacity gaps. Working on the principle that building resilience begins with an understanding of regional and local climate, the Meteorological Service Singapore established a Centre for Climate Research Singapore (CCRS) to build in-house capability on climate science and modelling. This is in addition to forging partnerships with international experts.

The climate change discussion is one of the strongest, if not the greatest factor shaping the development agenda. However, while numerous initiatives have been targeted to educate and raise awareness of other relevant stakeholders, there is still a great deal of work to be done. For example, in Lao PDR, while climate impacts are already felt by rural communities in the form of extreme weather events, climate change studies have not been communicated to them.

In a nutshell, the adaptation readiness varies by countries in SEA. Further, while adaptation readiness is not directly linked to a country's economic strength, it can be a defining factor which can either work for or against the said country's CCA efforts. This is because a country's economic strength highly correlates with its capability to allocate resources internally as well as its technical capability to access and mobilise external resources from various sources of finance. For purposes of this assessment, the following elements were applied to assess adaptation readiness: (1) policy, strategy, legal provision, and governance; (2) institutional arrangement; (3) integration and mainstreaming; (4) finance; (5) implementation capacity; (6) technology; (7) decision-making tools and institutionalisation; and (8) awareness, knowledge, and access to information.

Review of the national documents revealed that Brunei and Timor Leste are not yet ready to adapt to future impacts of climate change as they lack climate change strategy and action plan to support planning and implementation of adaptation measures. It should be noted however that efforts are already underway in both countries as indicated in their INDC and NAPA, respectively. Myanmar and Lao PDR can be said to have low adaptation readiness exacerbated by the countries' classification as LDC. Myanmar has formulated its National Adaptation Programme of Action (NAPA) and advancing process to undertake formulation of a National Adaptation Plan (NAP). Meanwhile, Salamanca and Nguyen (2016) stated that 'adaptation planning is very recent in Lao PDR'. The country has however mobilised various technical working groups and has established a national steering committee. Vietnam, while not falling under LDC group, also has low adaptation readiness as gleaned from challenges relating to policies, capacities, finance, and technologies indicated in their INDC.

Malaysia and Thailand, while both in the process of developing their national adaptation strategies, are perceived to have low to medium adaptation readiness. As Salamanca and Nguyen (2016) noted, these countries are the 'wait-and-see adaptors' which may be because they are classified as upper middle-income countries and have lower risk indices as compared to neighbouring SEA countries. This can however create problems for the two countries in the future especially with the high uncertainty factor involved in CCA. Singapore can be said to have medium adaptation readiness. While Singapore has an advantage economically, its inherent challenges linked to resource availability as well as lack of observable climate data hamper full implementation of CCA efforts.

As gleaned from this desk review, it can be said that Cambodia, Indonesia, and Philippines have medium to high adaptation readiness. What is common across these countries is that all have already established local financing mechanisms for CCA (e.g. climate trust funds). Further, Cambodia is the highest recipient of multilateral CCA financing in the sub-region. These countries also have plans and institutions in place to tackle CCA concerns. However, moving from planning to implementation as well as strengthening integration are concerns that these countries still need to address.

Considering this assessment, it can be said that while many adaptation actions have been prioritised and partially implemented, a number of factors underpinning smooth implementation can hinder the enhancement of said options. Some constraints include cost-effectiveness of adaptation initiatives, level of certainty of the risks involved, and competition with other development priorities. There are still many institutional, policy, capacity, and technical constraints in SEA that need to be addressed first before adaptation actions can be enhanced. These include proper implementation of climate policies and laws, improved coordination among relevant ministries, mechanisms to generate alternative sources of adaptation financing, economic priorities, access to information especially for local communities, and M&E system for CCA programmes and projects.

8.6 Attaining the Adaptation Vision

The incessant and intensifying impacts of extreme weather events suffered by the SEA region have contributed to the increasing concern of national governments to address climate change. Over the past decade, new institutions emerged and/or restructured existing institutions while policies, plans, and strategies were drawn and continuously updated following detailed scoping assessments and rigorous consultations with relevant stakeholders. However, implementation of these plans, to a large extent, boils down to access to adequate finance and technologies, proper coordination among the ministries and institutions. Based on the desk review and assessment, it has concluded that several areas of gaps need to be addressed for attaining the adaptation vision:

Create an enabling policy and organisational framework for ministries to coordinate and work together (horizontal integration) Coordination is defined as 'the organisation of the different elements of a complex body or activity so as to enable them to work together effectively'. The climate change agenda cuts across several issues concerning agriculture, forestry, water, energy, health, infrastructure, and social development. Hence, multi- and cross-sectoral involvement is non-negotiable. As discussed in the previous section, SEA countries recognise the need for such institutional arrangements.

Setting up of technical working groups and multi-sectoral steering committees is merely the initial step It is more important to ensure that these sub-groups are coordinating and functioning properly. Based on the challenges presented by SEA countries, there is a need to review mandates, roles, and functions of each institution involved to ensure that there is no overlap of responsibilities. Furthermore, relationships between and among these working groups need to be spelled out clearly to avoid turf wars. A designated entity needs to have the authority to do the overall coordination. The World Bank reported that 'execution and coordination of climate action are hindered by a lack of clarity in roles and responsibilities across institutions' in the Philippines. Fragmented support and limited decision-making opportunities has resulted in ineffectiveness in implementing climate agenda. This process will take time and continuous refinement before horizontal integration is achieved. As Indonesia's SNC put it, 'systematic and integrated efforts with a reliable strategy, as well as joint commitment and responsibility of various stakeholders are needed in mainstreaming climate change in national or local development agenda'.

Ensure vertical integration of climate change action plans As national-level plans are developed and finalised, the mechanism to cascade this down to the sub-national and local levels is imperative. Countries may have developed good action plans at the national level, however, when these are not cascaded to local governments, communities will not benefit from it. Cambodia's strategic plan acknowledges that guidelines on sub-national climate change planning must take off from lessons learned and outcomes of community-based pilot projects implemented in recent years. The document stressed the importance of identifying entry points for local governments to access climate change technical expertise from line ministries. ICLEI SEAS' experience in implementing various adaptation projects in selected local governments of the Philippines highlight the importance of working across various levels of government. For example, the national government can help provide cities with relevant climate data to come up with a more thorough assessment.

Continue capacity building efforts and develop review mechanisms Capacity of human resources to proactively respond to climate change impacts is positively correlated with the attainment of a country's adaptation vision. Unfortunately, review of the NCs revealed that there is a lack of capacities in the region to undertake adaptation planning to climate change and undertake actions. Capacity gaps are often evident in terms of collection, analysis, interpretation, and utilisation of climate data; climate modelling; and assessment of technologies. In the Philippines, the World Bank reported that financing gaps for knowledge and capacity development tend to hamper implementation progress. While there is recognition of the need to build capacity, the financial counterpart is absent. Such mismatch needs to be reviewed by decision-makers and appropriate actions should be in place to close this gap.

There is also a need to look at the impacts of capacity building activities implemented because as reported by Indonesia in its SNC, 'there is no system in place at present to record and monitor the outcome of the [capacity building] activities'. As such, there is no benchmark to indicate whether capacity has already improved. This can also apply to other SEA countries. Usually, learning events tend to be one-time and no follow-through activities are conducted. However, breaking away from the usual notion of capacity building will entail rethinking of how workshops and trainings are designed and delivered.

Strengthen capacities of actors to undertake adaptation financing Fundamental paradigm shift is also important in terms of the financing landscape for adaptation. Myanmar's Initial National Communication (2012) explained that there may be 'need to redirect investments and fund flows from all sources to more sustainable climate-friendly interventions. It has also stated need of scaling up of international private and public finance and optimisation of the funds allocation by spreading the risk across private and public investors'. In the Philippines, the People's Survival Fund has been created an attempt to fill in the adaptation financing gap targeting local governments. However, GIZ (2013) stated that capacity constraints at the local level can potentially hinder access and allocation of said fund. There is still limited capacity in terms of translating national policies into local level plans.

Tap local partners to communicate scientific climate change assessments to various stakeholders In terms of decision-making tools such as vulnerability assessments, numerous approaches are available and national-level assessments have mostly been conducted. However, results are not downscaled to a level that can be readily used and understood by vulnerable communities. As mentioned in the previous section, where capacities are available, these often remain at the national level and/or in urban centres and municipalities only. Additionally, the associated cost of doing a full-blown vulnerability assessment can be overwhelming especially for rural and small island communities as well as other far-flung areas. Communicating scientific and technical assessments can also be difficult and will require different learning modalities and approaches. Toward this end, the role of intermediaries including civil society organisations and NGOs become more apparent. National and local governments should continuously seek partners who can serve as dissemination arm of climate change information.

Build public awareness on climate change SEA countries have undertaken several information, education, and communication (IEC) activities as gleaned from the NCs. There are also efforts to integrate climate change in the formal education sector such as in the Philippines and Thailand. Timor Leste has indicated that it will explore strategy for IEC. However, like capacity building activities, efforts should be made to examine the effectiveness of IEC activities as well as appropriateness of IEC materials developed. Malaysia's SNC noted that 'while overall public awareness on climate change and environmental issues can be said to have grown, only limited evidence of this can be noted from changed behavioural patterns'. Similarly, Vietnam shared that communities' awareness of climate change tend to focus on negative impacts including extreme weather events but does not translate to greener practices and sustainable production and consumption with an end goal of building low-carbon and resilient cities.

Once country visions on climate change are laid out, the region's adaptation vision centres on building a resilient society. Over the past 10 years or so, climate change discussions have become more pronounced and intense in the region. Slowly, progress is made and the sub-region has taken a proactive stance to show the global community that adjustments are happening in SEA countries to address climate change. National and local governments alike continuously explore development pathways complementing efforts to minimise the adverse impacts of climate change.

The road to attaining the adaptation vision still entails facing a number of bumps and humps along the way. The constraints facing the region represent a complex web of factors with each one able to influence the other to some extent. For a region comprised mostly of developing countries, it is a daunting reality that adaptation continues to compete with other priorities such as health, education, and poverty reduction. SEA countries need to treat adaptation as an extension of sustainable development practices (Brömmelhörster 2009). It is only when adaptation is viewed through this lens that adaptation projects can become sustainable and effective.

References

Asia Pacific Adaptation Network & ICLEI Southeast Asia Secretariat (APAN- ICLEI) (2015) Climate change adaptation monitoring and evaluation in Southeast Asia. APAN website: http://www.asiapacificadapt.net/

Brömmelhörster J (2009) The economics of climate change in Southeast Asia: a regional review. Asian Development Bank. http://www.adb.org. Accessed 9 Mar 2015

Central Intelligence Agency (CIA) (2015) The World Factbook: East and Southeast Asia. CIA website. https://www.cia.gov/library/publications/the-world-factbook/wfbExt/region_eas.html. Accessed 10 Mar 2015

Climate Funds Update (2015) Adaptation. Climate Funds Update website. http://www.climate-fundsupdate.org/themes/adaptation. Accessed 10 Mar 2015

Climate Funds Update (2016) The data: climate finance recipients. Climate Funds Update website. http://www.climatefundsupdate.org/data. Accessed 6 Mar 2016

Global Environment Facility (GEF) (2008) Elements for an M&E Framework for Climate Change Adaptation Projects: lessons from GEF Climate Change Adaptation Projects. GEF Evaluation Office. http://betterevaluation.org/sites/default/files/GEF_EO_M_E_Framework_Adaptation%20%281%29.pdf. Accessed 10 Mar 2015

International Fund for Agricultural Development (IFAD) (2014) Comprehensive environment and climate change assessment in Viet Nam. IFAD website. http://www.ifad.org. Accessed 9 Mar 2015

Salamanca A, Nguyen H (2016) Climate change adaptation readiness in the ASEAN countries. Stockholm Environment Institute (SEI) Discussion Brief. http://www.sei-international.org/. Accessed 4 Mar 2016

UNFCCC (2015) NAPAs received by the secretariat. UNFCCC website. http://unfccc.int/adaptation/workstreams/national_adaptation_programmes_of_action/items/4585.php. Accessed 8 Mar 2015

UNFCCC (2016) UNFCCC submissions: intended nationally determined contributions. http://www4.unfccc.int/submissions/indc/Submission%20Pages/submissions.aspx. Accessed 24 Feb 2016

United Nations Economic and Social Commission for Asia and the Pacific (UNESCAP) (2014) Statistical Yearbook for Asia and the Pacific. Available at https://www.unescap.org/publications/statistical-yearbook-asia-and-pacific-2014

United Nations Framework Convention on Climate Change (UNFCCC) (2015) Non-Annex I National Communications. UNFCCC website. http://unfccc.int/national_reports/non-annex_i_natcom/items/2979.php. Accessed 8 Mar 2015

World Resources Institute (WRI) (2009) Continuum of adaptation activities: from development to climate change. WRI website. http://www.wri.org/resources/charts-graphs/continuum-adaptation-activities-development-climate-change. Accessed 10 Mar 2015

Yusra A (2013) APAN-SEAN CC Training Workshop building capacity on access to funds for CCA initiatives, 24–26 June 2013. Bangkok, Thailand

Yusuf A, Francisco H (2009) Climate change vulnerability mapping for Southeast Asia. EEAPSEA website. http://www.eepsea.org/pub. Accessed 8 Mar 2015

Chapter 9
Status of Climate Change Adaptation in Agriculture Sector in Asia

Rico C. Ancog, Mariliza V. Ticsay, and Clarissa D. Ruzol

Abstract Climate change adaptation in agriculture involves a systematic process of understanding, planning, implementation, and evaluation including structural and non-structural measures across various decision-making spheres. Generally, climate change adaptation in the agriculture sector in Asia is now primarily undertaken under the tenets of achieving food security and safety. The mode of implementation of climate change adaptation in agriculture varies across countries in Asia mainly because of the heterogeneous historical development of the region's (1) sectoral policies on the environment and natural resources; and (2) utilisation of the land and water resources that are closely knitted to cultural identity. The common vision for agriculture in Asia under a climate change scenario is resilience of the food production system in terms of (1) enhancement of crop yields to sustain the population demand; (2) economic stability; and (3) reduction in vulnerability of farming communities. Clearly, the differences among countries can be attributed to the varied understanding of climate change and its spatial and temporal implications as well as the country's level of economic growth. A trend among country reports suggests that climate change adaptation in agriculture is a combination of policy, research, and developmental discussions on food security and environmental governance. Overall, countries in Asia require access to sustainable financial mechanisms in order to support policy, research, and developmental projects on CCA in agriculture.

Keywords Asia · Climate change adaptation · Agriculture · Resilience · Adaptation status

R. C. Ancog (✉) · C. D. Ruzol
School of Environmental Science and Management, University of the Philippines Los Baños, Laguna, Philippines
e-mail: rcancog@up.edu.ph

M. V. Ticsay
Southeast Asian Regional Center for Graduate Study and Research in Agriculture (SEARCA), Manila, Philippines
e-mail: mvt@searca.org

M. Alam et al. (eds.), *Status of Climate Change Adaptation in Asia and the Pacific*, Springer Climate, https://doi.org/10.1007/978-3-319-99347-8_9

9.1 Introduction

Changes in the climatic conditions specifically temperature rise and variability in precipitation highly influence the productivity of agriculture sector in Asia. Climate change will primarily impact water resources and agriculture in Asia including crop yield, livestock, fishing, irrigation, pest and diseases, carbon fertilisation, food storage and distribution, and food accessibility and utilisation (Cruz et al. 2007). Climate scenarios reveal that changes in atmospheric CO_2 concentration, water availability, and temperature limit the predictability of crop production such as rice, wheat, and maize (Masutomi et al. 2009; Peng et al. 2004; Vermeulen et al. 2013). Though impacts of climate change on food productivity vary across Asia with favourable effects in the northernmost areas (Lal et al. 2001), developing countries and least developed countries in many of the sub-regions will experience decline (IPCC 2014; Nelson et al. 2009).

Climate change vulnerability of agriculture is a focus of interest in science, policy, and development agendas but the impacts on the components of the food production system, such as in food processing, transportation, consumption, and employment, have not been given similar emphasis (Behnassi et al. 2014). Approximately 83.68% of the world's GDP from agriculture is derived from the region. Almost half of Asia's economically active population are engaged in the agricultural sector that occupies 35.74% of its total land area (Table 9.1).

Non-climate stressors such as population increase, land use change, market failures, and policy and governance also aggravate the uncertainties and vulnerability to climate change impacts of agricultural sectors (Morton 2007; Vermeulen et al. 2013). Uncertainties may be derived from a country's exposure and sensitivity to climate change but impacts on agriculture have a clear direct relationship with economic stability. Economic losses due to impacts of climate change on agriculture particularly burden the livelihood of small-scale farming households. Moreover, adaptation in agriculture is crucial for countries where economic development is needed (Christiansen et al. 2011). This shows that climate change impacts will significantly alter current capacities of the region's food security and livelihood of small-scale farming communities. Hence, an assessment of the status of climate change adaptation in agriculture in Asia is needed to analyse current trends and identify ways to enhance capacities for addressing the future.

9.2 Agriculture Sector's Adaptation Objectives

A common vision among Asian countries is to improve food production (Fig. 9.1) in order to maintain food security and adequately supply the demands of the population. Economically sound management of water resources and food systems are the major concerns of climate change adaptation in agriculture in Asia. In countries capable, a more specific technology-driven reduction of vulnerability in the

Table 9.1 Agriculture in Asia, 2013

	Population			Land area			GDP		
Region	Total (millions)	Total economically active	% active in agriculture	Total (sq. km in thousands)	Agricultural (sq. km in thousands)	% Total	Total (US$ in billions)	Agriculture share (US$ in billions)	% Total
Central	83.33	40,062	18.81	4188.9	2983.47	71.22	446.4	40.42	9.06
North East	3075.92	1,034,352	49.62	27,262.04	7573.48	27.78	17,886.92	1115.76	6.24
South	1666.59	675,418	49.69	7516.6	3621.40	48.18	3410.99	481.08	14.10
South East	601.73	310,147	44.38	4320.6	1292.91	29.92	2338.78	260.43	11.14
Asia	5427.57	2,059,979	48.25	43,288.14	15,471.26	35.74	2,4083.09	1897.69	7.88
World	7124.54	3,434,318	38.72	134,324.7	51,043.39	38	75,592.94	2267.79	3
% World	76.18	59.98	–	32.23	30.31	–	31.86	83.68	–

Data Sources: World Bank (2013) and FAOSTAT (2013)

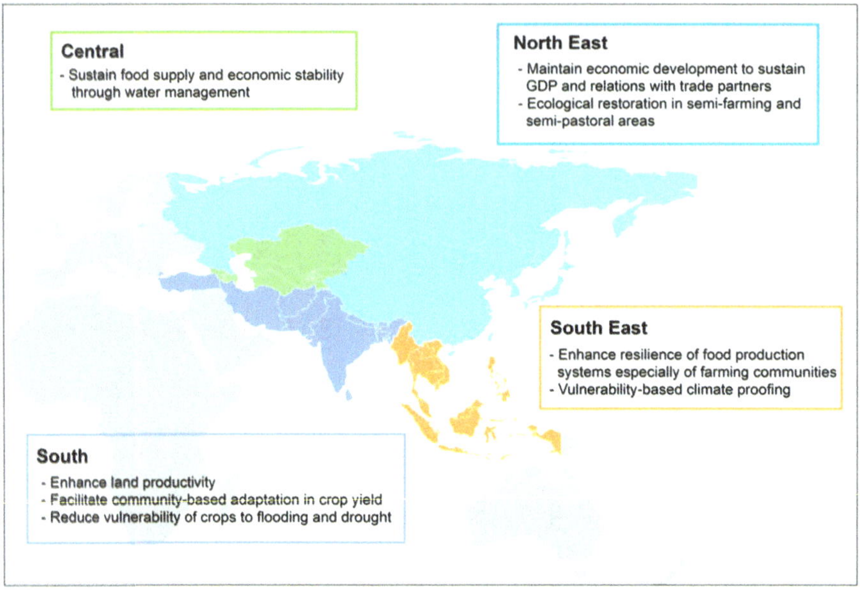

Fig. 9.1 Climate change adaptation objectives in agriculture in Asia

sub-national and community level has progressed to become a pivotal strategy of adaptation in agriculture. Differences among countries can be attributed to the varied understanding of climate change and its spatial and temporal implications as well as the country's level of economic growth.

Considering dynamic nature of the sector and complexities including double exposure of the sector, climate change adaptation in agriculture sector should place central emphasis on research and development to determine the set of climate change adaptation objectives and strategies. The dynamics and complexities arise from biophysical, government, economic, social, and application of technologies (Smithers and Blay-Palmer 2001) where technology innovation as a climate change adaptation strategy addresses the food production system and the society instigating technological, behavioural, and institutional change.

The IPCC reports, among other published literature, have presented innovations in agricultural research and development to cope with impacts of climate change. For instance, mid-term and long-term projections across different scenarios and modelling techniques attempt to increase the confidence of estimates to more accurate levels despite limited data. From the list of recommended adaptation measures from several studies, priority adaptation strategies are determined according to the assessment of current vulnerabilities of a country. Objectives of adaptation in Asia need to be discussed in a sub-regional level given the topographical, demographic, and cultural diversity compounded by the huge range of climate zones in the region. The significant vulnerabilities to climate change in each sub-region are briefly discussed in this section to provide the context for identifying the adaptation objectives.

9.2.1 Central Asia

One of the strongest indicators of the warming global temperature is permafrost degradation (Turral et al. 2011) such as in the Tien Shan mountain range in Central Asia (Marchenko et al. 2007; Narama et al. 2009). The potential increase in soil erosion from downstream runoff to the Aral Sea basin would likely affect agricultural productivity in adjacent areas (Lioubimtseva and Henebry 2009). Prioritising adaptation strategies for agriculture at the sub-regional level is an issue of adapting to water conflict and stress. In the northern areas, production could benefit from longer growing season and warmer winters while the southern portion of Central Asia could experience drought, desertification, and water scarcity (Hijoka et al. 2014). With more than 70% of the total land area of Central Asia classified as agricultural, countries are in unison that sustaining the food supply and maintaining economic stability through water management are key adaptation measures to climate change in the agricultural sector.

9.2.2 Northeast Asia

The North East sub-region of Asia comprises Mongolia, Japan, the Democratic People's Republic of Korea, the Republic of Korea, China, and the Russian Federation. Agriculture in the northernmost areas is described by semi-farming and semi-pastoral practices dominated by maize monoculture and livestock animals, namely sheep, goat, and cattle (Keio University's RCCCA 2012). Mongolia and parts of China are highly vulnerable to desertification. Modelled scenarios under changes in rainfall and temperature expect an increase in total crop production (measured in calories) in Japan and the Republic of Korea up to 2050, but decline is estimated to happen in China by as much as 10% (ADB 2013). In terms of rice yield, decrease in rice production in western Japan and eastern China is expected due to their high vulnerability to climate variabilities (Masutomi et al. 2009). Being dominated by economically developed and competitive nations, climate change adaptation in North East Asia is oriented to maintain economic development specifically to sustain GDP and the economic relations with trade partners. In contrast, Mongolia addresses the desertification problem by prioritising ecological restoration programmes and activities through the technological, educational, and financial assistance from countries such as Japan, China, European countries, and multilateral development banks.

9.2.3 South Asia

South Asia is bounded on the north by the Himalayan Mountain Range and on the south by the Indian Ocean. Across projected scenarios, the crop and livestock production in South Asia is expected to decline to as much as 30% by mid-twenty-first century (Parry et al. 2005; IFAD 2009). This could seriously threaten food security of more than one billion people (Elbehri et al. 2011). Decline in yield of desirable crops also encourages weed and pest proliferation (Ahmed and Suphachalasai 2014). In other areas of the sub-region, erratic patterns of monsoon and temperature trend are the major phenomenon of climate change. There is an increasing pressure to food stability especially in the poorest areas of South Asia where the level of chronic undernourishment is highest (Schmidhuber and Tubiello 2007). Surface runoff from glacial thawing is also a major concern in low-lying agricultural areas particularly in Nepal. The general adaptation objective in South Asia is therefore geared towards ensuring food security by enhancing land productivity, facilitating community-level adaptation, and reducing vulnerability of agricultural crops to flooding and drought in specific areas.

9.2.4 Southeast Asia

In the tropical region of Asia, there is an expected increase in rainfall variability and occurrence of extreme events such as drought and floods that strongly influence the agricultural sector in terms of yield and disaster losses especially in the rice production system. Droughts that are associated with El Niño Southern Oscillation years have caused massive crop failures, water shortages, and forest fires in the Philippines, Laos, and Indonesia (Cruz et al. 2007). It is estimated that agricultural production in Southeast Asia needs to increase by 60% by 2050 to satisfy the expected demand if the sub-region's population growth, income, and consumption continue at its present rate (FAO 2013). With these known hindrances to food security, Southeast Asia has progressed with the integration of climate change adaptation into other development planning such as in disaster risk reduction and green growth. Mainstreaming of this integrated policy in the sub-national level is also one of the common national framework agenda through community-based adaptation. Climate change adaptation in Southeast Asia aims to enhance resilience of food production systems as well as building resilience of agricultural communities through technology driven and vulnerability-based climate-proofing.

9.3 Adaptation Trend in Agriculture Sector

Adaptation begins by identifying climate change risks through vulnerability assessment. There are several approaches to vulnerability assessment but those particularly used in agriculture to determine risks include analysing livelihood security, poverty alleviation, disaster risk management, and land use planning. Scenario modelling, integrated remote sensing and geographic information system mapping and simulation, calculation of vulnerability indices, and economic valuation are common tools and techniques used to reduce uncertainties associated with vulnerability assessments. These methods are being improved across countries in Asia on an ongoing basis. Still, prevalence of complex uncertainties especially in long-term climatic conditions and highly dynamic socio-economic indicators are increasing the complexity of baseline assessment. Many countries in Asia are facing challenge of downscaling the global models of climate change scenarios at country and sub-national levels in order to provide context-specific measures.

Another key component of climate change adaptation in agriculture is the accessibility and availability of funding. Countries in Asia such as Nepal, Philippines, Indonesia, and Iran have specified domestic funding mechanisms by the national government allotted for autonomous adaptation strategies. Funding is also brimming with external opportunities from international organisations, academic universities, research institutions, and the developed countries. Examples of funding mechanisms for climate change adaptation are listed in Table 9.2.

The mode of implementation of climate change adaptation in agriculture varies across countries in Asia mainly because of the heterogeneous historical development of the regions, (1) sectoral policies on the environment and natural resources; and (2) utilisation of the land and water resources that are closely knitted to cultural identity. For example, community-based adaptation is being implemented in areas where there have been large investments in social capital. For example, in Nepal a people-initiated national network "The Federation of Community Forest Users" was formed. In other sub-regions such as in Southeast Asia, similar community-based adaptation is being implemented among indigenous groups and cultural communities. Fostering community engagement in climate risk reduction through strong linkages among non-government organisations helped address the communication of climate information to the community level in the Philippines, Indonesia, and Vietnam (van Aalst et al. 2008).

In the case of semi-arid and arid regions of Central Asia, transboundary management of water resources to combat drought and desertification is deemed to be the most effective way to address water conflict and stress. In particular, regional interaction and support from the international community are prerequisites to the enhancement of hydro-meteorological monitoring of snow and ice resources in Central Asia (IPCC 2014). Generally, the national government through the local government is mandated to take necessary actions. The state's obligation is further reiterated with the countries ratification to the UNFCCC that requires for a national communication report on climate change mitigation and adaptation. An emerging trend in climate change adaptation in agriculture is the development of co-benefits in the energy sector. For example, a project in Cambodia was funded by the Global Environmental Facility to conduct a pilot study on the use of agricultural residue biomass for sustainable energy solutions.

An important stage in the adaptation process is the monitoring and evaluation. Currently, only few countries in Asia have specified monitoring and evaluation guidelines of climate change adaptation interventions in agriculture or in any sector at all. While data is widely available for the monitoring and evaluation of climate change mitigation through national inventory of GHG emissions and carbon sequestration under the UNFCCC, in the case of adaptation, assessment of efficiency is

Table 9.2 Sources of funds for climate change adaptation

Fund	Description	Source
Adaptation Fund	Established under the Kyoto protocol of the UNFCCC and has committed US$ 331 million in 54 countries since 2010 to climate adaptation and resilience activities	https://www.adaptation-fund.org
Least Developed Countries Fund	Established to support a work programme to assist least developed country parties (LDCs) carry out, inter alia, the preparation and implementation of national adaptation programmes of action (NAPAs)	http://unfccc.int/cooperation_and_support/financial_mechanism/least_developed_country_fund/items/4723.php
Special Climate Change Fund	Established under the UNFCCC in 2001 to finance projects relating to: adaptation; technology transfer and capacity building; energy, transport, industry, agriculture, forestry and waste management; and economic diversification. This fund should complement other funding mechanisms for the implementation of the convention	http://unfccc.int/cooperation_and_support/financial_mechanism/special_climate_change_fund/items/3657.php
Global Climate Change Alliance	Established by the European Union in 2007 to strengthen dialogue and cooperation with developing countries, in particular LDCs and small island developing states (SIDS)	http://www.gcca.eu
Climate and Development Knowledge Network	Supports decision-makers in designing and delivering climate compatible development by combining research, advisory services, and knowledge management in support of locally owned and managed policy processes	http://cdkn.org
Global Facility for Disaster Reduction and Recovery	A global partnership that helps developing countries better understand and reduce their vulnerabilities to natural hazards and adapt to climate change. It provides grant financing, technical assistance, training, and knowledge sharing activities to mainstream disaster and climate risk management in policies and strategies. Managed by the World Bank, GFDRR is supported by 34 countries and 9 international organisations	https://www.gfdrr.org
International Climate Initiative	Finances climate change mitigation and adaptation and biodiversity projects in developing and newly industrialising countries, as well as in countries in transition. A key element of Germany's climate financing and the funding commitments in the framework of the convention on biological diversity	http://www.international-climate-initiative.com

focused on the ratio of the outputs (goods and services delivered) and the inputs (adaptation costs/ funding) of the programme (Brooks et al. 2013). The difficulty of quantifying the effects of adaptation is grounded upon the current uncertainties on climate change impacts in agriculture as a result of lack of data and the limitation of tools and models to aid decision-making.

Adaptation strategies for agriculture in Asia correspond to the climate change adaptation objectives (Fig. 9.1). Across sub-regions in Asia, countries invest in four common adaptation strategies. These are (a) water resource management, (b) development and use of climate-resilient crops, (c) insurance and incentive mechanisms for livelihood and food security, and (d) capacity building in the community and institutional level. Table 9.3 below provides a summary of climate change adaptation strategies for agriculture in Asia.

9.4 Gap and Areas of Adjustment

9.4.1 Policies, Strategies, Legal Provision, and Governance

Consistent among countries in Asia is the existence of a climate change policy or a related national report such as in the form of the National Communication to the UNFCCC, the National Adaptation Programme of Action (NAPA) for Least Developed Countries, and National Climate Change Adaptation Plans (Fig. 9.2). Most of the countries in Asia have prepared and outlined strategies in agriculture as a response to the UNFCCC's demand for National Communication. In the recently concluded Conference of Parties in Paris, 31 of the 38 countries in Asia submitted an Intended Nationally Determined Contributions (INDC) report to the UNFCCC. It is a new international agreement that publicly outlines the context of their national climate actions under a global framework towards a low-carbon and climate-resilient future.

Least developed countries were aided financially and technically through the formulation of NAPAs. Only seven countries in Asia have mainstreamed climate change adaptation into national policy, namely Japan, Korea, China, India, Sri Lanka, Indonesia, Malaysia, Philippines, Singapore, and Thailand, and process of integration is progressing in other countries in the region. In developed countries, such as Japan, adaptation initiatives in agriculture aim to provide assistance to less developed countries in design and implementation of their adaptation plans.

The current trend reflects the dependency of the legislation of policies and strategies primarily on the policy makers' motivation to pursue the international community's call to act. Generally, there is a wide disparity in the policy process between least developed, developing, and developed countries in Asia. There is an evident need to update and institutionalise climate change adaptation strategies in agriculture while integrating existing agricultural policies. Policies should be adaptive to climate uncertainties that are geared towards a transformational goal of institutional, behavioural, and attitudinal change in order to ensure better adaptation in the future.

Table 9.3 Summary of climate change adaptation strategies for agriculture in Asia

	Adaptation strategies	Source
East Asia		
Mongolia	Restoration of riparian zones; pastureland, and grazing management; development of sedentary livestock production technology; water management of mountain and steppe ecosystems	Keio University's RCCCA (2012) and Adaptation Fund (2011)
Japan, Republic of Korea, China, Mongolia	Consumer subsidy designed to restore food consumption in calories at the average level particularly from net imports; introduction of drought-tolerant varieties; increase in water use efficiency; agricultural input management; diversification of source of income (e.g. aquaculture and agroforestry); enhancement of pest and disease management; seasonal climate forecasting; selective migration	ADB (2013) and Easterling et al. (2007)
Central Asia		
Georgia	Development of emergency response plans for agriculture; innovative irrigation management and water application techniques; site-specific anti-erosion measures; establishment of information centres for farmers	Based on the 2015 INDC of Georgia
Kazakhstan	Integrated water resource management; agricultural crop diversification	Based on the 2009 SNC of Kazakhstan
Kyrgyzstan	Improve water resource management including irrigation efficiency, regulation of surface water flow, creation of water reserves, and effective water distribution; introduction of economic mechanisms that encourage activities of individual owners; introduction of incentives for water use efficiency	Based on the 2009 SNC of Kyrgyzstan
Tajikistan	Selection of drought-resistant and highly productive agricultural crops; zoning of land according to its vulnerability to erosion; soil protection measures such as land reclamation and crop rotation; rehabilitation of drain constructions, collectors, and pumping stations to combat swamping and salinisation; forest rehabilitation in regions prone to drought and wind erosion; systematic monitoring and forecasting of pasturelands; securing financial sustainability of households and agricultural crop and livestock insurance	Tajikistan National Action Plan for Climate Change Mitigation (2003) and SNC (2008)
Turkmenistan	Improved efficiency of water use and storage; introduction of less-moisture consuming crops; enhancing alternative sources of water such as slightly saline drainage water, groundwater, and treated industrial and municipal wastewater; strengthening of policy and institutional capacity, communal management system for water delivery	Based on the 2002 National Environmental Action and 2011 Adaptation Fund of Turkmenistan

(continued)

Table 9.3 (continued)

	Adaptation strategies	Source
Uzbekistan	Improvement of irrigation and drainage infrastructure; preserving and increasing of fish resources; develop institutional and technical capacity for drought management, soil conservation and moisture retention, and early warning systems	Based on the 2009 SNC and 2015 Adaptation Fund of Uzbekistan
South Asia		
Afghanistan	Eco-agriculture and climate friendly irrigation technology transfer	Based on the 2015 INDC of Afghanistan
Bangladesh	Increase mechanisation	Based on the 2015 INDC of Bangladesh
Nepal	Risk identification; community-based integrated management; improved management of livelihood assets of food insecure households; early warning system for glacial lake outburst flood; improve local, district, and national capacity to plan, implement, and monitor adaptation	Based on the 2010 NAPA and 2015 Adaptation Fund of Nepal
Bhutan	Diversification of agricultural crops; performance of religious rituals (rimdo); soil and land management technology to reduce soil erosion; initiation of crop and livestock insurance scheme; shift to intensive greenhouse agriculture/aquaculture; upgrade storage facilities; national capacity to develop and implement emergency response to agricultural pest and disease outbreaks/epidemics	Lal et al. (2001), Davis and Li (2013), and Wangdi et al. (2013)
India	Development of drought- and pest-resistant crop varieties through biotechnology; enhancing soil conservation and water use efficiency; community-based information sharing and dissemination of local agro-meteorological data; introduction of financial mechanisms to support farmers to invest and adopt technologies; strengthening of current agricultural and weather insurance mechanisms; climate-resilient farm designs; mobile soil-testing laboratories	Based on the 2009 National Action Plan on Climate Change and 2015 Adaptation Fund of India
Southeast Asia		
Cambodia	Develop forest restoration (i.e. establishment of multi-use forests) and conservation agriculture; institutional capacity, awareness raising and upscaling of eco-agriculture interventions; building sea dikes and scaling up of climate-smart farming systems	Based on the 2012 Adaptation Fund and 2015 INDC of Cambodia
Indonesia	Land expansion of farming and aquaculture areas; adjustment of food production system; repair and development of infrastructure; acceleration of food diversification; development of innovative and adaptive technology; development of information and communication systems	Based on the 2013 National Action Plan for Climate Change Adaptation (RAN-API) of Indonesia

(continued)

Table 9.3 (continued)

	Adaptation strategies	Source
Myanmar	Use of climate-resilient crop varieties, ex/in situ conservation of plant genetic resources, vulnerability reduction of rural and subsistence farmers	Based on the 2012 National Action Plan on Climate Change of Myanmar
Philippines	Enhancement of site-specific knowledge on agricultural vulnerability; support researches on technologies and the dissemination of results; creation of knowledge base management; integration of CCA and DRR in national and local levels as well as in the education curricula; scaling up of best practices; monitoring and evaluation of CCA and DRR programs; implementation of risk transfer and social protection mechanisms; capacity building in the government, communities, and industry	National Climate Change Action Plan of the Philippines 2011–2028
Lao PDR	Promotion of alternative sources of livelihood for farmers affected by disasters; shifting to short-duration paddy and other cash crops; strengthening of technical capacities in the national and local level; crop and livestock animal diversification; promotion of integrated pest management; empowerment of farmer organisations; development of bank erosion protection	Based on the 2009 NAPA of Lao PDR
Timor-Leste	Develop sustainable land management including increase in soil fertility, prevention of erosion, and promotion of permanent agriculture; improve legal framework specifically on livestock production	Based on the 2010 NAPA of Timor-Lester

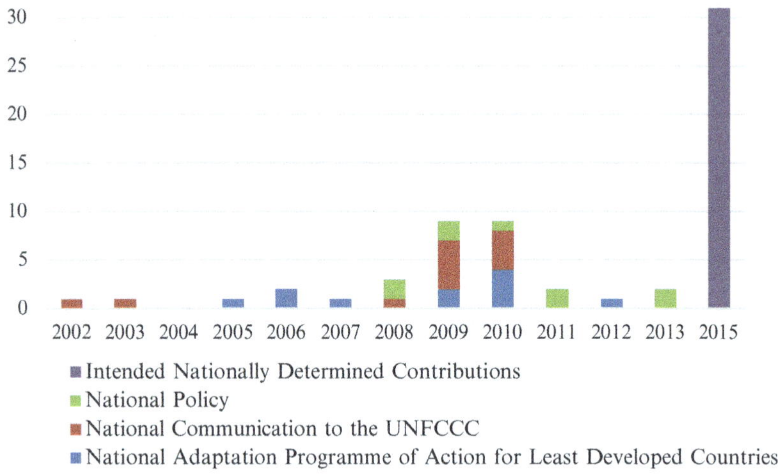

Fig. 9.2 Most recent reports on climate change adaptation from each country in Asia

9.4.2 Institutional Arrangement

The vulnerability of the agricultural sector is also influenced by the institutional framework (Ahmed and Suphachalasai 2014). Two types of institutional arrangement are currently influencing climate change adaptation in agriculture in Asia: (1) government and (2) non-government. A well-placed institutional framework of the government can bridge the communication and coordination gap between sectoral ministries and government agencies, which is often lacking in many agricultural countries in Asia. There are existing institutional arrangements in countries such as in the Philippines, China, Mongolia, Japan, among others. In the Philippines, for instance, the Climate Change Commission is composed of the Climate Change Office, the National Panel of Technical Experts, and the Advisory Board comprising the Secretaries of different departments including the Department of Agriculture, presidents of leagues of local governments, and representatives from the academe, business sector, and non-government organisations. However, many other countries have not set out clear institutional arrangement that outline functions to legislate, oversight, implement, monitor, and evaluate the strategies for climate change adaptation, specifically in agriculture.

In addition, the role and opportunity for transboundary or regional cooperation, such as in the case of water management, is often being recommended, but remains not fully explored. Many research institutions—local, national, regional, and global—are especially active in the scientific development of new technologies, development of crop varieties, vulnerability assessment, disaster risk reduction and management, and adaptation that directly influence decision-makers including those at the community level. Some of these institutions in Asia include the System for Analysis, Research and Training, the Institute for Global Environmental Strategies, the Watershed Organisation Trust, and the Southeast Asian Regional Center for Graduate Study and Research in Agriculture. Government and non-government research institutions are both crucial in the development, adoption, and ensuring sustainability of adaptation in agriculture. Synchronisation of their interventions can hold legitimacy in the political, social, cultural, and scientific contexts of the vulnerable agricultural areas. For most of the countries in Asia that lack national policies on climate change adaptation, implementation of strategies in agriculture is hindered by the weak coordination among government agencies and non-government research institutions. This also puts constraints on the accessibility of resources including financial, technical, and human that may otherwise be efficiently mobilised.

9.4.3 Integration and Mainstreaming

Mainstreaming climate change into national development framework to the local government level is a commonly perceived strategy as long-term solution across countries in Asia particularly in East and Southeast Asia. Integration however has proven to be much more difficult to attain because of bureaucratic and political barriers. Social capital between and among the implementing body and the recipient community is weak. This is manifested by the weak accountability and transparency as well as the lack or inadequate dialogues between the authorities and the farmers.

Another aspect that hinders integration and mainstreaming in CCA in agriculture in Asia is the ambiguous delineation of functions of many of the policies on climate change and sectoral policies. Although localising adaptation at the village level has been recognised as pivotal, resources and capacities are still centralised and the means to avail the services often fail to be communicated to the most vulnerable. A strong information, education, and communication campaign should be put into place taking into consideration the principles of risk communication. The appropriation of the concepts from technical studies in climate change impacts on agriculture to the local language and culture is urgently needed for farmers to comprehend and value adaptation. This is also likely to merge the apparent gap between socio-cultural acceptability and technology-driven adaptation. Aside from the fragmental support given to the necessary strengthening of social capital and recognition of cultural diversity, strategic procurement of funds is also essential to mobilise resources and knowledge for agricultural adaptation.

9.4.4 Finance

The cost of adaptation is far less than the cost of damages of climate change impacts, which makes it only rational for governments to invest in adaptation in agriculture. In the countries of Southeast Asia, namely Indonesia, Philippines, Thailand, and Vietnam, investment in drought- and heat-resistant crops and sea walls would only entail 0.2% of their combined GDP that could potentially avoid damages amounting to 1.9% of the combined GDP by 2100, on an annual basis (ADB 2009). However, the potential adaptation options particularly those involving engineering and infrastructural adjustments such as in irrigation and wastewater treatment require outsourcing of funds especially in developing and least developed countries in Asia.

A large share of funding of adaptation programs in agriculture in Asia comes from international sources. International financing of climate change adaptation in all sectors in Asia is estimated to be at around US$ 50.3 million, next to Africa with as much as US$ 118.6 million dispersed (www.climatefundsupdate.org). Currently, there has been a total of US$ 134.8 million approved funding for projects on climate change adaptation strategies in agriculture in Asia from multilateral agencies (Figs. 9.3 and 9.4). Cambodia has received a total of US$ 40.7 million to implement projects on climate-resilient agricultural practices, services, and infrastructures, strengthening of adaptive capacity of rural communities, and technology transfer on the use of agricultural residue biomass for sustainable energy.

Availability of funds dedicated to climate change is globally ubiquitous but are more inclined to research and development in sustainable alternative energy. Climate finance resources for agriculture, on the other hand, are primarily being disbursed to countries most vulnerable to climate change such as in Bangladesh,

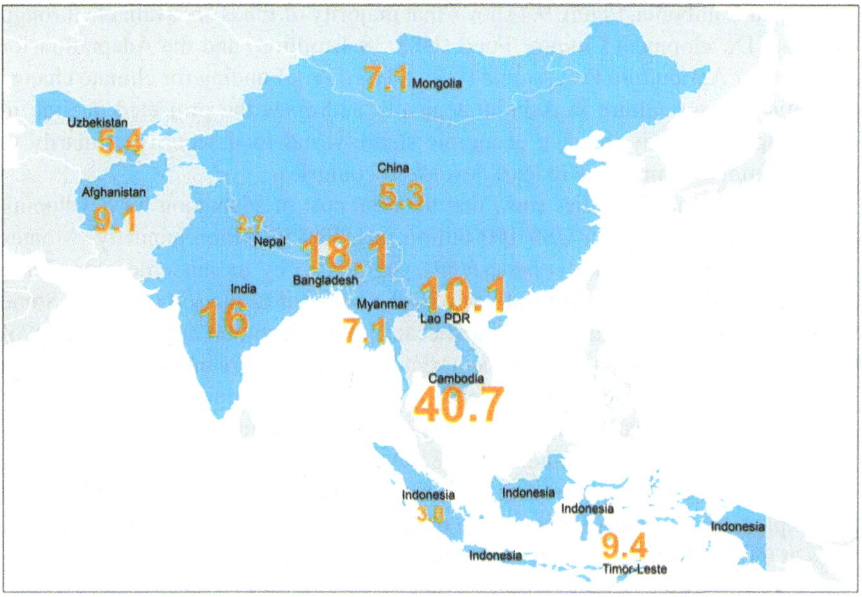

Fig. 9.3 Approved Climate Finance for Agriculture in Asia in US$ millions (Adaptation Fund 2015; Climate Funds Update 2015)

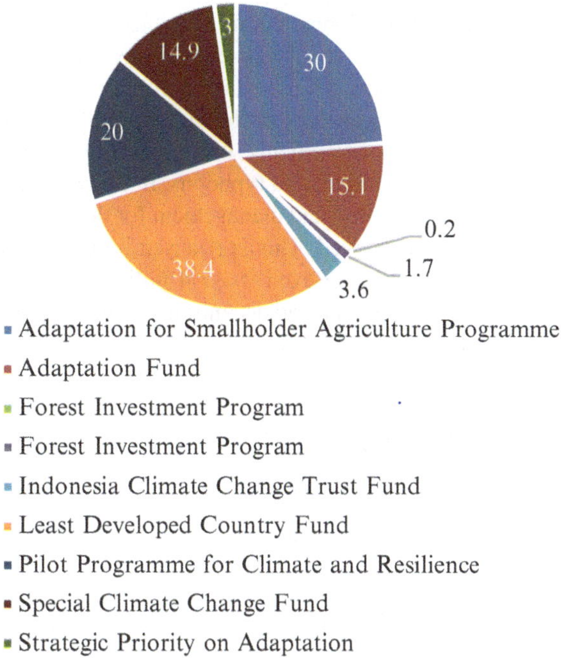

Fig. 9.4 Sources of Funds for Climate Change Adaptation in Agriculture in Asia (US$ millions), (Adaptation Fund 2015; Climate Funds Update 2015)

Nepal, and Cambodia. Figure 9.4 shows that majority of funds are available through the Least Development Country Fund (US$ 38.4 million) and the Adaptation for Smallholder Agriculture Programme (US$ 30 million). Funding for climate change adaptation in agriculture in Asia strategically addresses the projected decline in food supply that may threaten economic stability and food security primarily in small farming communities in least developed countries.

On the other hand, studies show that the total cost of adaptation for developing countries may range from US$ 100 billion to US$ 450 billion annually (Montes 2012). Insufficient funding is persistently augmented by the underdeveloped and biased accessibility and mobilisation from the national to the village level. Some funding agencies are difficult to attract because of the apparent huge uncertainty of the impacts of climate change on the agricultural sector. Streamlining of financial mechanisms from the national as well as international funding institutions to the local authorities is necessary to encourage autonomous and contextualised farm-level adaptation. The assessment of the economic cost of adaptation should be strengthened to give concrete evidence for policy making. Finally, financing should also explore mechanisms for incentive programs and payments for ecosystem services at the local level as means to sustainably finance local adaptation.

9.4.5 *Implementation Capacity*

The success of implementation is directly proportional to the investments in individual and organisational capacities across scale of decision-making. There are excellent adaptation options available for agriculture sector but failure in implementation has been a prevailing problem. The insufficient capacity to implement is especially hindering the resilience of agricultural systems to climate-related disasters such as extreme rainfall and typhoons. In South and South East Asian countries, the post-disaster response and management for the loss in crop and livestock production is significantly delayed. This lag caused by poor implementation capacity and not economically efficient even if good policies were put into place. There are also cases where centralised and decentralised implementation may both yield beneficial results when authorities are capacitated and supported by the political leader.

9.4.6 *Technology*

The level of technological advancement is often associated with economic competency of a country measured in GDP and per capita income since it requires capital investment in high-end tools. Technological tools for climate adaptation can be categorised into hard technologies, soft technologies, and organisational technologies. Hard technologies involve infrastructural and physical adjustments usually at the farm level. Adoption of hard technologies and its integration into the farmer's decision-making is the desired outcome for climate adaptation.

The farmer's decision-making could also be aided by soft technologies, a practice more common in climate change adaptation in agriculture in Asia. Remote sensing techniques and mobile applications are being applied for precision farming to aid farmers' decision-making. Mobile phone applications for agriculture, or more popularly known as "mobile farming", targets to enhance productivity and yield, access market and value chains, and improve provision of public service to farmers (World Bank 2011). Mobile farming applications for agriculture are already being used in many countries in Asia for community-based agro-climate monitoring (Quiang et al. 2011). This technology merges scientific knowledge with local knowledge such that farmers receive advice and guidelines on crop information, soil properties, pests and diseases, microclimate data, and irrigation and drainage. Farming communities particularly in India and Japan use this application to increase the resiliency of agricultural land against perturbations such as drought or erratic rainfall. This technology can be used for early warnings and advisory broadcast to mitigate agricultural risks such as occurrence and spread of diseases and market price volatility (World Bank 2011), which are expected to be exacerbated by climate change.

Another significant example of soft technology for climate change adaptation in agriculture is knowledge management that varies across different sub-regions in Asia. In South East Asia, the Knowledge Center on Climate Change (KC3) of the Southeast Asian Regional Center for Graduate Study and Research in Agriculture (SEARCA) (http://www.climatechange.searca.org) provides comprehensive resources focusing on climate change and risk management in agriculture in South East Asia. Another online knowledge database is the UN FAO Climpag (http://www.fao.org/nr/climpag/about_en.asp), an open portal on climate change, climate variability, and agriculture under the general concept of food security. The site offers free access to international data on climate monitoring, global climate grids, climate data and maps, spatial interpolation, and other publications. Open access technical data such as these are crucial for local governments that are mandated to localise and mainstream the national climate change framework strategy such as in the Philippines.

In countries where there is very limited local climate data, opportunities for research exchange and training programs from developed countries are being encouraged. Mongolia, for example, has been a recipient of researchers from Japan, Korea, China, and from other parts of Europe who give trainings on the applications of remote sensing and geographic information system to climate change vulnerability assessment particularly on desertification. Research institutions, though they are actively engaged, are located and often delimit their services to specific areas and countries of interests. While these trainings particularly educate organisations about available soft technologies, knowledge about how other institutions in other countries address climate change in agriculture is also learned by experience. Such learning could be acquired from organisational technologies for climate change adaptation in agriculture, which focus on mainstreaming, capacity building, and enhancement of management skills in the managerial and institutional level. Examples of organisational technologies for climate change adaptation in Asia are farmer field schools, community-based agricultural extensions, forest user groups, and water user organisations.

The government agencies, ministries, and research institutions responsible for developing technologies and knowledge, however, are inadequate to keep pace with the increasing risk of agriculture by climate change. The irreducible uncertainties in country-level effects of climate variabilities in agricultural production question the accuracy of modelling and assessment tools. Because of this high uncertainty, financing mechanisms commit lesser investments in assessing the impacts of climate change to agriculture than in other research and development sectors such as clean energy and climate change mitigation. In addition, while the climate change phenomena have gained considerable attention in the policy and academic discourse since the first world climate conference in the 1970s, there are few specialists whose expertise involves downscaling of sophisticated modelling analysis in many countries in Asia. These gaps should be minimised to attain the ultimate goal of research and development, that is, to provide empirical, significantly accurate, and integrated climate scenarios in agriculture to aid decision-making.

9.4.7 Decision-Making Tool and Institutionalisation

The decision-making process involves the coordination between government agencies, technical experts, and representatives of civil society. Representativeness of stakeholders is highly encouraged for the preparation of national policies, NAPAs, and the national communication to the UNFCCC. Decision-making is being coordinated by a council on climate change or, more often, the government agency responsible for natural resource and environmental management. This framework in decision-making favours the potential benefits from the plurality of perspectives and accessible resources but demands commitment in participation and communication amounting to a high transaction cost. Moreover, government agencies are expected to perform their mandated roles and functions but current roles on climate change adaptation in agriculture are vaguely defined and contestable. Integration of sectoral agencies for decision-making addressing a common goal also proves to be highly complex where contradicting priorities may emanate (i.e. prioritising food security by expanding agricultural land as opposed to forest conservation to improve carbon stocks).

9.4.8 Awareness and Access to Information

Lack of awareness is one of the most frequently mentioned potential barriers in implementation in country reports and national policies. There is a wide difference in the access to information among countries in Asia critically abated by climate change impacts. Awareness of climate change impacts to crop yield, soil quality, and water availability are learned by farmers and other relevant interest groups from various available media publicised by the government, non-government groups, or private sectors. First-hand observation of impacts is also highly regarded to cultivate a sense of value to adapt to climate change among communities of farmers and users. For example, maintaining indigenous knowledge about agricultural practices is being promoted in Nepal and Bhutan.

In other cases, however, current adaptation in agriculture is hindered by the continuous unsustainable utilisation of water and land resources. This is further burdened by the limited knowledge about climate change impacts of the members of the government. Educating government officials about the implications of the impacts of climate change to agriculture will catalyse legislation of appropriate actions and strengthen the assistance to farm-level adaptation. Awareness raising could be improved by the collaboration of the government with non-government organisations through emphasising the benefits of adaptation in the medium- and long term. Malaysia in particular specifies that education and awareness raising can be done formally and informally but desired outcomes should be clearly identified during the planning stage.

9.5 Readiness for Enhanced Adaptation Actions

Climate change adaptation in agriculture involves a systematic process of understanding, planning, implementation, and evaluation. Climate change adaptation plans in Asia include structural and non-structural measures for effective land use planning, reduction in vulnerability, economic diversification, early warning systems, integrated water resource management, cropland management, and restoration of degraded soils (Smith et al. 2007; IPCC 2014). Generally, implementation of adaptation strategies in agriculture is limited by financial, institutional, technical, organisational, and lack of public awareness about climate change, but varies in different countries (Table 9.3).

The most frequently mentioned potential barrier to the implementation of adaptation strategies in agriculture is the insufficiency of data on country-level climate change vulnerabilities, impacts and methods to increase the confidence level of projected scenarios. World climate models present that climate change will negatively impact crop yield and water availability in most areas in Asia, but uncertainties in projections hinder institutionalisation of policy. In Russia, for example, according to its Third National Communication in 2002, there are anticipated benefits from climate change in plant cultivation through increase in land suitable for agriculture, longer vegetative periods, and warmer temperature during winter.

In Bangladesh, understanding and technical knowledge about climate change impacts on agriculture are very much there and adequate which resulted in engagement of different actors at almost all levels to address impacts on agriculture. Efforts include introduction of climate-resilient crop varieties to deal with increased salinity in coastal areas, submergible crop varieties to deal with flood. Maintaining food production and ensuring food security are main drivers behind its continued efforts to make the sector resilient to climate change. National development plan has already mainstreamed climate change which calls for action by each sector to integrate climate change in project design and implementation. However, considering sheer size of population and declining trend of cultivable land, a country requires scaling up of its efforts and adaptation interventions in agriculture sector and need to move beyond crop agriculture paddy. It also requires strong leadership of the government to integrate at systemic level, which can be explained by the absence of a climate change component in sectoral policies. Decision-making of the government and farmers could be aided by enhanced informed awareness campaigns about the known impacts of climate change on the food production system.

Related to the lack of technical knowledge and public awareness, the review of most recent reports of countries in Asia reveals that there are organisational difficulties in the actualisation of climate change adaptation strategies in agriculture. Mainstreaming of national strategies for climate change adaptation in agriculture can be aided by information and communication technologies (ICTs) such as in mobile farming in Japan and India. Mobile farming creates a community and a network linking farmers, farmer's organisations, buyers, local governments, and the scientific community. It offers flexibility and access to opportunities to coordinate with multiple stakeholders for innovative agricultural practices, sustainable land management, good e-governance, and financial stability. The emergence of new ICTs and their application to agriculture help decision-makers to respond during emergency situations. Consistent with the goals of the society to become sustainable, ICT tools should be promoted in other countries especially in the community level.

Financial mechanisms are viewed to be adequate by most countries in Asia and only 25% mentioned about financial constraints (Table 9.4). Least developed nations reiterated the need for more funding to projects of climate change adaptation in agriculture. Reallocation of climate finances is being mobilised through multilateral funding from developed countries. Adaptation trend shows that priority areas for most vulnerable countries in Asia are funding for adaptation interventions but a more comprehensive cost-benefit analysis is being demanded to persuade more investments in the agricultural sector. Overall progress in climate change adaptation in agriculture in Asia would be gradual and supported by the development of science and technology to deliver more accurate and credible projections which is likely to result in more investment in the sector.

Monitoring and evaluation (M&E) of climate change adaptation in agriculture in Asia is underdeveloped. This is because of the indefinite observable impacts of climate variabilities through time and space leaving policy makers uncertain about future policy options. However, there are efforts to develop M&E frameworks to track results of adaptation. Indonesia emphasises the necessary selection of key M&E indicators limiting the complexity the M&E framework while taking into account all relevant sectors through a participatory approach. Stakeholder consultation and strength of a participatory approach in adaptation highly depend on the country's social capital across levels of decision-making and implementation.

Table 9.4 Policy and reports on climate change adaptation in agriculture most recently published by countries in Asia

Sub-region	Country	Report	Year published	Potential barriers in implementation					
				Financial	Institutional	Technical	Organisational	Public awareness	Conflict
North East	Japan	Approaches to CCA	2010	–	–	–	–	–	–
	Democratic People's Republic of Korea	NA	NA	NA	NA	NA	NA	NA	NA
	Mongolia	SNC	2010		X	X	X	X	
	People's republic of China	National plan for addressing CC	2013			X	X	X	
	Russian Federation	TNC	2002			X			
	Republic of Korea	CCA plan	2011		X	X	X		
Central	Armenia	TNC	2015			X			
	Azerbaijan	SNC	2010			X		X	
	Georgia	SNC	2009		X	X	X	X	
	Kazakhstan	SNC	2009			X	X	X	
	Kyrgyzstan	SNC	2009		X	X	X		
	Tajikistan	SNC	2008			X		X	
	Turkmenistan	SNC	2009		X	X	X	X	
	Uzbekistan	SNC	2009		X	X	X	X	

(continued)

Table 9.4 (continued)

Sub-region	Country	Report	Year published	Potential barriers in implementation					
				Financial	Institutional	Technical	Organisational	Public awareness	Conflict
South	Afghanistan	NAPA	2009	X	X	X	X	X	X
	Bangladesh	NAPA	2005		X	X		X	
	Bhutan	NAPA	2006					X	
	India	National action plan on CC	2009		X		X		
	Islamic Republic of Iran	SNC	2010			X			
	Maldives	NAPA	2007	X	X	X	X	X	
	Nepal	NAPA	2010	X	X	X	X		
	Pakistan	INC	2003	X	X	X	X	X	
	Sri Lanka	National strategy on CC	2011	X	X	X	X		
South East	Brunei Darussalam	NA	NA	NA	NA	NA	NA	NA	NA
	Cambodia	NAPA	2006	X	X	X	X		
	Indonesia	National action plan for CCA	2013	–	–	–	–	–	–
	Lao PDR	NAPA	2009	X	X	X	X	X	
	Malaysia	National policy on CC	2009		X	X	X	X	
	Myanmar	NAPA	2012	X	X	X	X	X	
	Philippines	National framework strategy on CC	2010			X	X	X	
	Singapore	National CC strategy	2008	–	–	–	–	–	–
	Thailand	National strategy on CC	2008		X	X	X	X	
	Timor-Leste	NAPA	2010	X	X	X	X		
% Total mentioned				**28.125**	**46.875**	**78.125**	**62.5**	**53.125**	**3.125**

NAPA National Adaptation Programme of Action for Least Developed Countries; *INC* Initial National Communication to the UNFCCC; *SNC* Second National Communication to the UNFCCC; *TNC* Third National Communication to the UNFCCC; *NA* not available; – not specified

Review of available reports and data indicates that countries lagging behind the design and implementation of adaptation strategies in agriculture can be complemented by regional coordination between developed, developing, and least developed countries. Asia as a region has the capacity to address and reduce future vulnerabilities of climate change in agriculture. Countries, on the other hand, need to strengthen national and local institutional and organisational capacities primarily by raising awareness about climate change impacts.

9.6 Attaining the Adaptation Vision

The common vision for agriculture in Asia under a climate change scenario is resilience of the food production system in terms of (1) enhancement of crop yields to sustain the population demand; (2) economic stability; and (3) reduction in vulnerability of farming communities. Adaptation strategies in agriculture are closely associated with water resource management due to the projected impacts of changes in rainfall patterns on agro-climatic conditions and people's utilisation of water resources.

A trend among country reports suggests that climate change adaptation in agriculture is a combination of policy, research, and developmental discussions on food security and environmental governance. These contemporary societal issues require a holistic and scientific understanding of uncertainties at the regional, national, and local level. Technical interpretation of climate scenarios alone could not translate into successful practices if adoption of agricultural technologies overly simplifies the local context. While the standardisation of methodological assessments and the preparation and implementation of climate policies in agriculture are significant to systematically compare differences and recognise similarities, applications to specific localities should be done with informed caution. It is therefore an undeniable fact that a shared commitment among multiple stakeholders across levels of government, the academe, non-government organisations, and the farming community is pivotal to achieve climate resilience in agriculture. A framework for achieving the identified adaptation vision in agriculture in Asia is illustrated in Fig. 9.5. Since countries in Asia emphasise about the importance of awareness raising in CCA in agriculture, the framework promotes behavioural and attitudinal change along with institutional change as a consequence of enhanced awareness.

Limited climate data and the dynamic socio-economic influences on agriculture have directed research more on vulnerability assessment and less on confidently recommending potential adaptation options. Uncertainty can be minimised by comparing various methodological frameworks to come up with a mixed-method approach for a context-specific implementation of vulnerability assessment (Ancog et al. 2016). In addition, an integrated understanding of key components of climate change in agriculture relevant to the end users (e.g. the farmers, the community, and indigenous peoples) can be facilitated by supplementing and validating technical knowledge with the farmers' world views and observations of the changing climate.

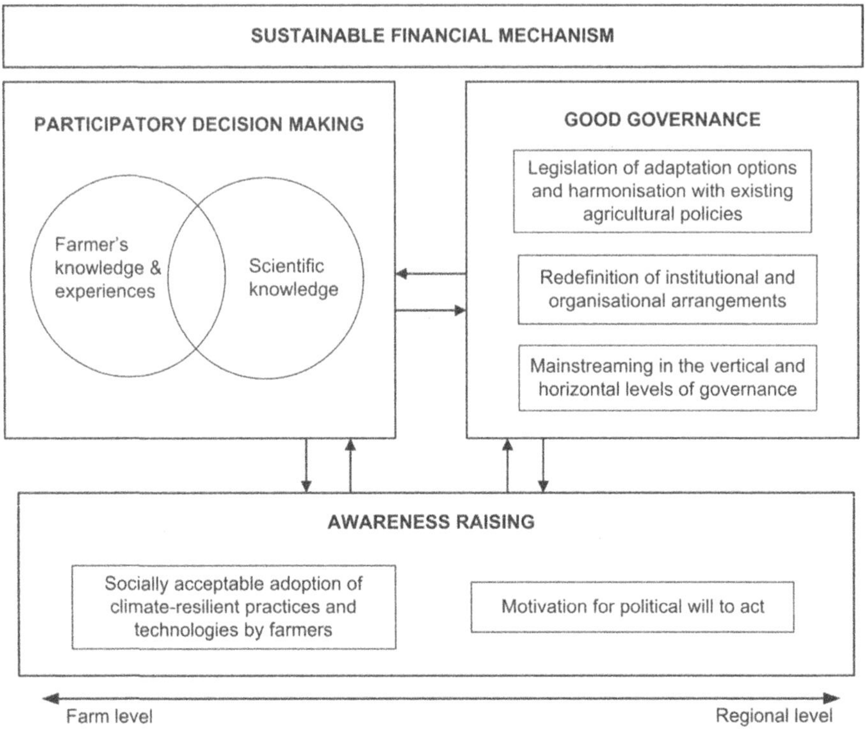

Fig. 9.5 A framework for enhancing climate change adaptation in agriculture in Asia

The development of new technologies and tools to accurately measure impacts of climate change on the different components of the food production system is crucial for evidence-based policy making. Integration of local and scientific knowledge provides an opportunity to raise in-depth awareness about the different perspectives on climate change among "experts", farmers, and policy makers.

Policy and governance mechanisms should redefine institutional arrangements according to existing agricultural policies while integrating climate change adaptation to national planning. This task is highly complex and difficult to achieve because CCA in agriculture demands adaptability and flexibility while policies are conceived to be established firmly on official rules and sanctions. An in-depth awareness about the complexities of climate change adaptation in agriculture by policy makers should give them a better perspective of the costs and benefits of adaptation.

On the other hand, research should promote innovation of climate-resilient agricultural practices and technologies that are socially acceptable to the recipient users. Mainstreaming of adaptation options in the community level through culturally appropriate communication (i.e. accurate translation of concepts to the vernacular language) would ensure sustainability of community-based implementation, monitoring, and evaluation.

Finally, countries in Asia require access to sustainable financial mechanisms in order to support policy, research, and developmental projects on CCA in agriculture. Outsourced financial resources can be complemented by incentive mechanisms to further minimise transaction costs and achieve cost-effectiveness in implementation, monitoring, and evaluation.

References

Adaptation Fund (2015) http://adaptation-fund.org. Accessed 12 Jun 2015

ADB (2009) The economics of climate change in southeast asia: a regional review. Asian Development Bank, Mandaluyong City

ADB (2013) Economics of climate change in East Asia. Asian Development Bank, Mandaluyong City

Ahmed M, Suphachalasai S (2014) Assessing the costs of climate change and adaptation in South Asia. Asian Development Bank, Mandaluyong City

Ancog R, Rebancos C, Sumalde Z (2016) Levels and determinants of vulnerability of two indigenous communities in the Philippines. Int J Clim Change Strateg Manage 8(2):154. https://doi.org/10.1108/IJCCSM-09-2014-0100

Behnassi M, Syomiti Muteng'e M, Ramachandran G, Shelat K (eds) (2014) Vulnerability of agriculture, water, and fisheries to climate change: towards sustainable adaptation strategies. Springer, Dordrecht. https://doi.org/10.1007/978-94-017-8962-2

Brooks N, Anderson S, Burton I, Fisher S, Rai N, Tellam I (2013) An operational framework for tracking adaptation and measuring development (TAMD), IIED Climate Change Working Paper No. 5. IIED

Christiansen L, Demery L, Kuhl J (2011) The (evolving) role of agriculture in poverty reduction—an empirical perspective. J Dev Econ 96(2):239–254

Climate Funds Update (2015) http://www.climatefundsupdate.org/. Accessed 12 Jun 2015

Cruz R, Harasawa H, Lal M, Wu S, Anokhin Y, Punsalmaa B, 2007. Asia. In: Parry M, Canziani O, Palutikof J, van der Linden P, Hanson C (eds) Climate change 2007: impacts, adaptation and vulnerability. Contribution of Working Group II to the Fourth Assessment Report of the Intergovernmental Panel on Climate Change, Cambridge University Press, Cambridge, p 469–506

Davis M, Li L (2013) Understanding the policy contexts for mainstreaming climate change in Bhutan and Nepal: a synthesis. Regional Climate Change Adaptation Knowledge Platform for Asia, Partner Report Series No. 10. Stockholm Environment Institute, Bangkok

Easterling W, Aggarwal P, Batima P, Brander K, Erda L, Howden S (2007) Food, fibre and forest products. In: Parry M, Canziani O, Palutikof J, van der Linden P, Hanson C (eds) Climate change 2007: impacts, adaptation and vulnerability. Contribution of Working Group II to the Fourth Assessment Report of the Intergovernmental Panel on Climate Change. Cambridge University Press, Cambridge, pp 273–313

Elbehri A, Genest A, Burfisher M (2011) Global action on climate change in agriculture: linkages to food security, markets and trade policies in developing countries. Food and Agriculture Organisation of the United Nations, Rome

FAO (2013) Climate-smart agriculture sourcebook. Food and Agriculture Organisation of the United Nations, Rome

FAOSTAT (2013) Food and Agriculture Organisation of the United Nations. http://faostat3.fao.org/. Accessed 2 Dec 2015

Hijoka Y, Lin E, Pereira J, Corlett R, Cui X, Insarov G (2014) Asia. In: Barros V, Field C, Dokken D, Mastrandrea M, Mach K, Bilir T (eds) Climate change 2014: impacts, adaptation, and vulnerability, part B: regional aspects. Contribution of Working Group II to the Fifth Assessment

Report of the Intergovernmental Panel on Climate Change. Cambridge University Press, New York, pp 1327–1370
IFAD (2009) Climate change impacts: South Asia. Retrieved from The International Fund for Agricultural Development (IFAD). http://www.ifad.org/events/apr09/impact/south.pdf. Accessed 12 Jun 2015
IPCC (2014) Summary for policymakers. In: Field C, Barros V, Dokken D, Mach K, Mastrandrea M, Bilir T (eds) Climate change 2014: impacts, adaptation, and vulnerability. Part A: global and sectoral aspects. Contribution of Working Group II to the Fifth Assessment Report of the Intergovernmental Panel on Climate Change. Cambridge University Press, New York, pp 1–32
Keio University's RCCCA Proceedings Report (2012) Educational Programme on a semi-arid environment susceptible to climate change in North East Asia, Horqin Sandy Lnd, Inner Mongolia, China. Hayama, Japan. Institute for Global Environmental Strategies, 25 August–3 September 2012
Lal M, Harasawa H, Murdiyarso D (2001) Asia. In: McCarthy J, Canziani O, Leary N, Dokken D, White K (eds) Climate change 2001: impacts, adaptation and vulnerability. Contribution of Working Group II to the Third Assessment Report of the Intergovernmental Panel on Climate Change. Cambridge University Press, New York, pp 535–581
Lioubimtseva E, Henebry G (2009) Climate and environmental change in arid Central Asia: impacts, vulnerability, and adaptations. J Arid Environ 73(11):963–977
Marchenko S, Gorbunov A, Romanovsky V (2007) Permafrost warming in the Tien Shan Mountains, Central Asia. Glob Planet Chang 56:311–327
Masutomi Y, Takahashi K, Harasawa H, Matsuoka Y (2009) Impact assessment of climate change on rice production in Asia in comprehensive consideration of process/parameter uncertainty in general circulation models. Agric Ecosyst Environ 131:281–291
Montes M (2012) Understanding the long term finance needs of developing countries, South Centre: presentation to the first workshop on long term finance. http://unfccc.int/files/cooperation_support/financial_mechanism/long-term_finance/application/pdf/montes_9_july_2012.pdf. Accessed 2 Jul 2015
Morton J (2007) The impact of climate changes on smallholder and subsistence agriculture. PNAS 1054(50):19680–19685
Narama C, Kaab A, Duishonakunov M, Abdrakhmatov K (2009) Spatial variability of recent glacier area changes in the Tien Shan Mountains, Central Asia, using (~1970), Landsat (~2000), and ALOS (~2007) satellite data. Glob Planet Chang 71:42–54
Nelson G, Rosegrant M, Koo J, Robertson R, Sulser T, Zhu T (2009) Climate change impacts on agriculture and costs of adaptation. International Food Policy Research Institute, Washington, DC
Parry M, Rosenzweig C, Livermore M (2005) Climate change, global food supply and risk of hunger. Phil Trans R Soc B 360:2125–2138
Peng S, Huang J, Sheehy J, Laza R, Visperas R, Zhong X (2004) Rice yields decline with higher night temperature from global warming. PNAS 101:9971–9975
Quiang CZ, Kuek SC, Dymond A, Esselaar S (2011) Mobile applications for agriculture and rural development. ICT Sector Unit-World Bank http://siteresourcesworldbankorg/INFORMATIONANDCOMMUNICATIONANDTECHNOLOGIES/Resources/MobileApplications_for_ARDpdf. Accessed 30 Oct 2015
Schmidhuber J, Tubiello F (2007) Global food security under climate change. PNAS 104(50): 19703–19708
Smith P, Martino D, Cai Z, Gwary D, Janzen H, Kumar P (2007) Agriculture. In: Metz B, Davidson O, Bosch P, Dave R, Meyer L (eds) Climate change 2007: Mitigation, Contribution of Working Group III to the Fourth Assessment Report of the Intergovernmental Panel on Climate Change. Cambridge University Press, New York
Smithers J, Blay-Palmer A (2001) Technology innovation as a strategy for climate adaptation in agriculture. Appl Geogr 21:175–197

Turral H, Burke J, Faures J (2011) Climate change, water and food security. Food and Agriculture Organisation of the Nations, Rome

van Aalst M, Cannon T, Burton I (2008) Community level adaptation to climate change: the potential role of participatory community risk assessment. Glob Environ Chang 18:165–179. https://doi.org/10.1016/j.gloenvcha.2007.06.002

Vermeulen S, Challinor A, Thornton P, Campbell B, Eriyagama N, Vervoort J (2013) Addressing uncertainty in adaptation planning for agriculture. PNAS 10(21):8357–8362

Wangdi T, Lhedup P, Wangdi N (2013) An analysis of forestry policy, acts and rule of Bhutan to mainstream climate change adaptation. Regional Climate Change Adaptation Knowledge Platform for Asia, Partner Report Series no. 13. Stockholm Environment Institute, Bangkok

World Bank (2011) ICT IN AGRICULTURE: connecting smallholders to knowledge, networks, and institutions, November 2011

World Bank (2013) Turn down the heat: climate extremes, regional impacts, and the case for resilience. Washington, DC.

Chapter 10
Adaptation in Mountain Agriculture: Food Security in the Hindu-Kush Himalayan (HKH) Region

Abid Hussain, Bidhubhusan Mahapatra, and Golam Rasul

Abstract Mountain agriculture is more vulnerable to adverse impacts of climate change as it heavily relies on rainfall. Although the impact of the climate change on agriculture and food security has become a serious concern, so far limited efforts have been made to understand how climate change impacts food security of the mountain communities of the HKH region, how poor households adapt to changing conditions, and what options are available to facilitate better adaptation. The nature and causes of agriculture and food security in mountains is quite different from the plains and require a specific set of policy measures. This chapter aims to provide a better understanding of the vulnerability of mountain agriculture and food security to climate change risks. It analyses the appropriateness of existing policies and programmes for mountains, policy and local level adaptation trends, areas of adjustments, and possible adaptation options such as climate smart agriculture, rainwater harvesting technologies, solar energy for irrigation, and better crop choices based on agro-ecological potential of specific areas conducive to improved food and livelihood security in the mountains. One of key questions that arises in the chapter is "How can the poor and marginalized mountain farmers be included in the adaptation process?"

Keywords Hindu-Kush Himalaya · Adaptation · Mountain · Agriculture · Food security · Livelihoods

10.1 Introduction

The Hindu-Kush Himalayan (HKH) region consists of vast mountain ranges covering a large geographical area (3,441,719 km^2) of eight countries: Afghanistan, Bangladesh, Bhutan, China, India, Myanmar, Nepal, and Pakistan (Fig. 10.1). It is rich in biodiversity and natural resources including ecosystem and ecosystem services and is the basis for livelihoods of around 240 million people living in this

A. Hussain (✉) · B. Mahapatra · G. Rasul
International Centre for Integrated Mountain Development (ICIMOD), Kathmandu, Nepal
e-mail: Abid.Hussain@icimod.org

M. Alam et al. (eds.), *Status of Climate Change Adaptation in Asia and the Pacific*, Springer Climate, https://doi.org/10.1007/978-3-319-99347-8_10

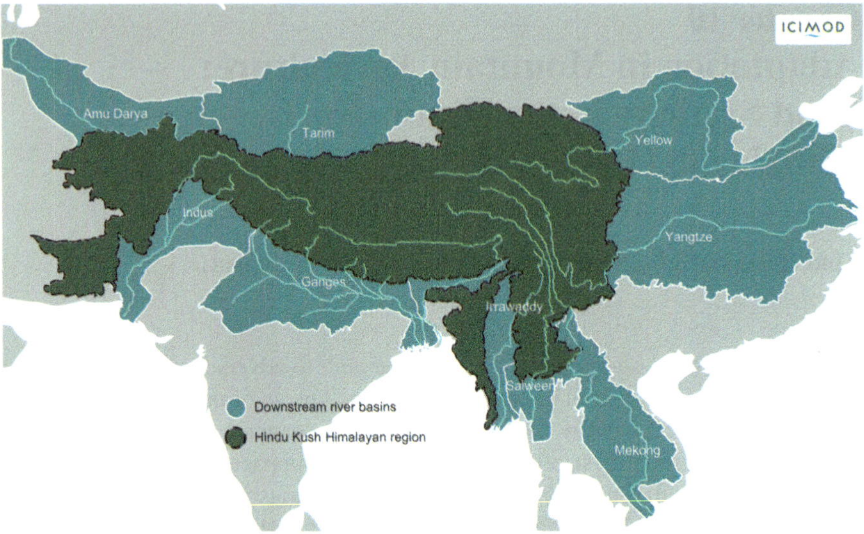

Fig. 10.1 Hindu-Kush Himalayan region

region. The HKH region is the source of river basins, which provide water to 1.9 billion people, one-fourth of the world's population (Rasul et al. 2018).

People living in the HKH region are highly vulnerable to food insecurity because of low productivity, subsistence economies, constraints of terrain, poor infrastructure, limited access to markets, physical isolation, vulnerability to natural risks, and high cost of food production and transportation (Tiwari and Joshi 2012; Giribabu 2013; Rasul et al. 2014). Worldwide, almost 842 million people are facing chronic hunger and 51% are from six HKH countries, i.e. Bangladesh, China, India, Myanmar, Nepal, and Pakistan (FAO 2013). Rising impacts of climate change (CC) have further added to the problem in both upstream and downstream areas due to temperature changes and frequent floods, severity and fluctuation of seasons, and unusual changes in precipitation. Climate changes occurring in mountain areas are also prompting severe impacts on downstream areas through increasing pressure on natural resources, production systems, and competition for food, accommodation, and income opportunities (ICIMOD 2008; Rasul et al. 2014). In South Asia, increased use of nitrogen fertilizers and manure to meet demands for food arising from rapid population growth contributes to increasing trend of greenhouse gases emission (IPCC 2007). Seasonal mean rainfall shows inter-decadal variability, noticeably a declining trend with more frequent deficit monsoons under regional in-homogeneities. It is also projected that climate change will affect food security by the middle of the twenty-first century, with the largest numbers of food-insecure people located in South Asia (IPCC 2014). It adds to the concerns of future variability of climate and its impacts in the region particularly in the mountain areas.

In the HKH countries, around 70% of the population lives and works in rural areas; the majority depends on agriculture for their livelihoods (World Bank 2015).

In the mountain ranges, the proportion of agriculture-dependent population is much higher than that of national average. Mountain farmers mainly practise rain-fed agriculture which is highly vulnerable to CC. Climate change is impacting particularly the current water scenarios, resulting in high concern about food security in the HKH. The past trends and the projected changes in temperature and precipitation in the HKK clearly show that the temperature and water availability is changing faster in the HKH region than other regions. Ensuring food security has become a big challenge in the face of such changes, given the other bio-physical and economic constrains in the mountains. As in Nepal, for instance, food insecurity has become a concern in Mid- and Far-Western Hill and Mountain districts (e.g. Khotang). Likewise, in mountain areas of Pakistan particularly in Federally Administered Tribal Areas, Baluchistan, Gilgit-Baltistan, and Khyber-Pakhtunkhwa, nearly 60% people are food insecure (FSA 2009).

Climate change has negatively affected the agriculture and food security in the HKH region. Recent evidences (ICIMOD 2014) have shown that climate change has resulted in reduced productivity of agriculture and increased incidences of floods and droughts. In addition to crops, climatic changes have also induced severe impacts on livestock through degradation of rangelands, pastures, and forests. In recent years, CC has led to a depletion of the natural resource base across the HKH region (Rasul et al. 2014). This has resulted in a significant loss of ecosystem services, particularly in terms of soil nutrients, water, and biomass, and resultant decline in food productivity (Tiwari 2000). The reduced productivity has affected food security and increased the overall vulnerability of mountain people, as they have become more dependent on food from outside and thus exposed to market fluctuations. Climate change is adding new challenges in terms of water availability and changing temperature conditions. Farming-households in HKH region practise mixed farming systems including the cultivation of cereals, fruits, and vegetables, and raising of different kinds of livestock. More than 80% farm-household's own livestock to support their livelihoods because crops are more vulnerable to CC impacts (ICIMOD 2015).

In the fragile mountain ecosystems of HKH, in addition to topographical and socioeconomic factors, food security and livelihoods of the people are also affected by CC impacts such as degradation of vegetation cover, deterioration of soil, and a reduction in livestock productivity, further impoverishing pastoral communities. It leads to a decline in food security of the mountain areas. Some evidences on mountain food security in the HKH region are presented in Table 10.1.

The region needs to continue to build the resilience of the agriculture sector to reduce CC risks. Climate change projections indicate that there will be increased water availability from higher rainfall and enhanced runoff from glaciers and snowmelt (Ministry of Environment 2010). Incidences of glacial lake outburst floods (GLOFs) are also increasing in the HKH mountain ranges. However, due to greater variability in the weather, some mountain ranges are also facing drought. In Baluchistan province of Pakistan, for instance, the drought spell (1998–2003) affected nearly 2 million acres of cultivable land and 9.3 million livestock and led to the death of 1.76 million livestock; nearly 80% of apple orchards were destroyed.

Table 10.1 Food (in)security in the Hindu-Kush Himalayan (HKH): some examples

Country	Indicators		Mountains	Plains	National average	Remarks
India	Dietary energy intake (kcal/day/capita)	Rural	2098	–	2147	Mountain statistics are estimated for states of Arunachal Pradesh, Assam, Manipur, Meghalaya, Nagaland, Sikkim, and Tripura
		Urban	2092	–	2123	
	Protein intake (g/day/capita)	Rural	56.4	–	59.3	
		Urban	57.1	–	58.8	
Nepal	Proportion of food insecure population (%)		59.5 (mountain) 52.8 (hills)	47.9	50.8	–
Pakistan	Proportion of food insecure population (%)		60.7	40.2	48.6	All districts of Pakistan taken into account
	Food deficit/surplus (%)		−50	+18	+9	Surplus = + Deficit = −

Sources: Giribabu (2013), MOHP Nepal (2012), FSA (2009), and Hussain (2010)

Food prices rose, and the food security and livelihoods of nearly two million mountain people were affected, resulting in reduced food consumption and the migration of people from drought-affected areas to areas downstream (Shafiq and Kakar 2006).

GLOFs caused heavy damages in Chitral district particularly in Mustuj, Lot Koh, and the Kalash valleys of Pakistan. Around 40 people lost their lives and thousands of houses were swept away. Floods also swept away major roads, power stations, irrigation and water systems, leaving an estimated 250,000 people stranded. About 70% of water supply channels have been destroyed and people have been left with no water for drinking or irrigation. In addition, standing crops and hundreds of livestock were affected severely. In the affected valleys, floods have caused some long-term damages to food security and livelihoods of the local people through damaging micro-irrigation systems and eroding small agricultural lands. The floods also washed away power stations and cut off electricity supplies, plunging the entire district into darkness. Phone services stopped working leading to a breakdown in communication with villages in upper Chitral, including the Yarkhun and Laspur valleys (DAWN 2015; Thethirdpole.net 2015).

GLOFs in Northern Pakistan: A Case Study

Glacial lakes pose a growing danger in the northern mountain areas of Pakistan. Chitral, the northernmost largest district of the Khyber-Pakhtunkhwa province, is home to about 542 glaciers covering nearly 13% of the district. It is estimated that about 40% of the district's population is vulnerable to the potential threat of GLOFs. In the recent past, the Bindo-Gol glacier lake, northwest of Chitral town, at an altitude of about 4572 m, had burst its banks. Reshun glacier in Reshun valley (4100 m) and Oxhoor glacier (4300 m) in Gharam Chashma had also burst, causing havoc in Karimabad town (Thethirdpole.net 2015).

(continued)

GLOFs in July 2015 together with unexpected heavily rainfalls resulted in flash floods. Chitral was the most affected district as glaciers are melting due to global warming and become destabilised. Unexpected heavy rainfall accelerated the speed of already melting glaciers due to temperature rise. A heat wave in Chitral during June and the first week of July—with temperatures soaring to 42 °C—caused glaciers to melt and form lakes. These lakes swelled and broke their banks when sudden heavy rain descended on the mountains. The warmer rainwater in comparison to the ice caused the ice to melt faster. The water accumulated at the base of the glaciers, creating ponds underneath. When the capacity was filled, GLOFs took place. Sometimes, the lakes (caused by glacial melt) are visible and can be monitored, but sometimes they are hidden under the surface of the ice and can't be monitored (DAWN 2015; Thethirdpole.net 2015).

Agriculture is the most important sector as far as food security of mountain communities is concerned. To make agriculture more resilient to CC, the local people and the institutions in HKH countries need to adapt to the new, more variable weather patterns that are emerging. This will require a series of actions ranging from climate smart practices, adjustments in infrastructure, improved water harvesting, and the adoption of improved cropping and farming systems, including new varieties that can adapt to different weather conditions. At the same time, there is a need for improved coordination and synergies between various institutions. In institutional arrangements for implementing CCA strategies and measures, local institutions such as local public institutions, civil society institutions, and private institutions are very important in implementation of adaptation strategies at local level.

10.2 Adaptation Objectives of Mountain Agriculture

The HKH countries are making progress towards framing instruments for better adaptation to CC particularly in the agriculture sector. These countries have incorporated climate change adaptation (CCA) for agriculture sector in their policies or national adaptation programmes. However, robust strategies for CCA specific to mountains are yet to be formulated (Table 10.2). Those HKH countries which have both mountainous and plain areas have established general objectives for CCA without complete and clear set of strategies and measures for mountain areas according to the level of their vulnerability and agro-ecological potential. For example, China has aimed to improve water management in mountains, but other measures, i.e. technology, resilient crop varieties, and livestock breeds, for improving productivity and food security in mountains are not clearly formulated for mountain ranges. Pakistan has formulated a good set of mountain-specific measures. Myanmar

Table 10.2 Objectives of HKH countries for climate change adaptation in agriculture

Country	General objectives and policy measures	Mountain-specific measures
China	• Improving agricultural infrastructures • Adjustment of agricultural structure and cropping systems • Promoting stress-resistant crop varieties and livestock breeds • Preventing aggravation of grassland desertification • Strengthening research and development of new agricultural technologies	• Accelerating the construction of water collection and utilization engineering in hill and mountain areas
Pakistan	• Establishing climate change impact assessment models • Improving inputs, irrigation and energy efficiency, adoption of solar energy, and crop productivity • Strengthening institutions on climate change science • Improving natural resource management • Promotion of livestock and horticulture • Establishing risk management systems for improving resilience crops and livestock	• Preventing climate-induced crop damages through introducing cold and drought-resistant short duration high yielding cereal crops • Introducing new feedstock technology for cattle and livestock suited to drier, harsher climates at high altitudes • Protecting agricultural terraces from extreme soil erosion, wind, hailstorm, and snowstorm-related damage • Promoting the use of gravity drip irrigation and hydraulic ram pumps in the mountains areas of Pakistan
India	• Improving resilience of Indian agriculture to heat and water-stress • Inclusion of indigenous knowledge in adaptation • Improving productivity of rain-fed agriculture	• Starting management measures for sustaining and safeguarding the Himalayan glaciers and mountain ecosystems
Bangladesh	• Developing climate change resilient cropping systems through new tolerant crop varieties of flooding, drought, and salinity • Promoting fisheries and livestock	• Specific strategies for Chittagong hill tracks are not revealed
Nepal	• Enhancing the climate adaptation and resilience capacity for optimum utilization of natural resources • Establishing Climate Change Centre as an effective technical institution	• Monitoring of Glacial Lake Outburst Flood (GLOF) for disaster risk reduction (likely to benefit agriculture) • Promoting green technology and climate smart practices in the utilization of natural resource for agriculture.
Bhutan	• Promoting community-based forest fire management and prevention, improved efficiency in irrigation and change cropping patterns, and promotion of agro-forestry to reduce soil erosion	• Almost the whole geographical area of Bhutan is mountainous or hilly

(continued)

Table 10.2 (continued)

Country	General objectives and policy measures	Mountain-specific measures
Myanmar	• Enhancing rice production through farm mechanization and breeding new rice varieties to ensure food security in areas most vulnerable to climate change	• Reducing CC vulnerability and increasing the resilience of rural and subsistence farmers particularly in dry and hilly zones
Afghanistan	• Improving resilience and productivity of agriculture through drought-resistant crop varieties and livestock breeds, post-harvest supporting services, emergency supplies to vulnerable communities, input supplies and promotion of horticulture, terracing, and agro-forestry	• Objectives for mountain areas are not specified except the promotion of terracing to reduce soil and water erosion. It may be assumed that general strategies apply to mountains also because most of the geographical area of Afghanistan is mountainous

Source: China National Climate Change Programme (2007), National Climate Change Policy of Pakistan (2012), National Action Plan on Climate Change of India, Bangladesh Climate Change Strategy and Action Plan (BCCSAP) & National Adaption Programme of Action (NAPA) (2009), Climate Change Policy of Nepal (2011), National Adaptation Program of Action of Bhutan (2006), National Adaptation Program of Action to Climate Change (2012) of Myanmar, National Adaptation Programme of Action for Climate Change (NAPA) of Afghanistan (2009)

has also aimed to reduce the vulnerability of mountain zones through promotion of climate-resilient crops. However, they have not aimed to improve technology, infrastructure, conservation, and integration of non-crop agricultural enterprises such as livestock and poultry, in the hilly areas. India and Nepal have also not differentiated the CCA measures for plain and mountain areas.

10.3 Adaptation Trends

10.3.1 Status of Policies and Programmes

Out of the eight HKH countries, five countries except Pakistan, China, and India have submitted NAPAs to the UNFCCC. National policies and/or adaptation programmes of all HKH countries reveal that the national steps and policy measures are showing very positive change towards better adjustment in the agriculture sector. However, they need to further strengthen their strategies and measures for better adaptation in the mountain agriculture to achieve sustainable food security. As a response to decision of Durban Conference of Parties to Framework Convention on Climate Change and invitation to Parties to initiate or intensify domestic action, all HKH countries also communicated their Intended Nationally Determined Contribution (INDC) in advance before the twenty-first session of the Conference of the Parties (COP).

China's National CC Programme clearly mentions agriculture as the priority sector for adaptation to CC (Government of People's Republic of China 2007). For CCA in agriculture, China aims to improve agricultural infrastructures to cope with the water-stress situation arising from CC. They have also planned to promote adjustment of the agricultural structure and cropping systems, stress-resistant crop varieties and livestock. Among other adaptation measures, they have focused on preventing aggravation of grassland desertification and strengthening research and development of new technologies in the areas of photosynthesis, biological nitrogen fixation, bio-technology, prevention of diseases and pests, stress resistance, and precision agriculture. China's National CC Programme does not clearly spell out mountain-specific strategies for CCA in agriculture except accelerating the construction of water collection structures and promoting utilization engineering (e.g. improved irrigation techniques) in hilly and mountainous areas. China's INDC also reiterate the focus on improving the construction of water conservation facilities for farmlands, vigorously developing water-saving agricultural irrigation and cultivating heat-resistant and drought-resistant crops (GoPRC 2015).

Pakistan has made significant progress towards framing policy instruments for CCA in mountain agriculture. National Climate Change Policy (NCCP) of Pakistan (Government of Pakistan 2012) sets clear policy measures for enhancing adaptations in mountain agriculture. It has focused on preventing climate-induced crop damages through the introduction of cold- and drought-resistant short duration high yielding cereal crops, introduction of new feedstock technology for cattle and livestock suited to drier, harsher climates at high altitudes, protecting agricultural terraces from extreme soil erosion, wind, hailstorm- and snowstorm-related damage, and promoting the use of gravity drip irrigation and hydraulic ram pumps[1] in the mountains areas. Pakistan's INDC are rooted in Vision 2025 which is aligned with the country's continued commitment to the issue of climate change and its impacts on agriculture, energy, water, and other sectors (Government of Pakistan (GoP) 2015a, b).

In India, the National Action Plan (NAP) on CC (Government of India 2008) has been devised to make Indian agriculture more resilient to CC. It aims to start management measures for sustaining and safeguarding the Himalayan glaciers and mountain ecosystems, but does not outline mountain-specific measures of CCA in agriculture. In general, the NAP focuses on new varieties of crops with higher resilience to heat, and alternative cropping patterns capable of withstanding extremes of weather, drought, floods, and variable moisture availability. India's INDC is national in scope and does not reveal mountain-specific adaptation strategies in agriculture.

Bangladesh developed the Climate Change Strategy and Action Plan (BCCSAP) in 2008, which was revised in 2009. Bangladesh also submitted its National Adaption Programme of Action (NAPA) to UNFCCC in 2005 which was updated in 2009. The BCCSAP specifically establishes action plan for adaptation in agriculture

[1]A hydraulic ram pump is a cyclic water pump powered by hydropower. It takes in water at one "hydraulic head" (pressure) and higher flow rate, and outputs water at a higher hydraulic head and lower flow rate.

under the domain of food security, social protection, and health. The action plan puts an emphasis on developing CC resilient cropping systems through agricultural research to develop crop varieties, which are tolerant to flooding, drought, and salinity, and based on the indigenous and other varieties suited to the needs of resource poor farmers. Promotion of fisheries and livestock is also included in the plan to provide more flexibility to adaptation practices for national food security (Government of the People's Republic of Bangladesh 2009). The action plan does not provide specific strategies for the Chittagong hill tracks. Moreover, INDC is also national in scope, and do not reveal any plans for hilly areas (Government of the People's Republic of Bangladesh 2015).

The Climate Change Policy (CCP) of Nepal (Government of Nepal 2011) reveals that water resources, agriculture, forestry, biodiversity, physical infrastructures, public health, tourism, and livelihood are highly vulnerable to CC. It focuses on green technology, climate smart practices in the utilization of natural resource, and strengthening of institutions in all ecological belts such as plains, hills, and mountains. In high mountains, monitoring of GLOF for disaster risk reduction has also been emphasized. The uniqueness of this policy is that it initiates the community-based local adaptation actions in ecological belts (plains, hills, and mountains) as mentioned in the NAPA. Nepal has prepared a National Framework for Local Adaptation Plan for Action (LAPA) with the twin objectives of implementing adaptation actions, and integrating CC into local development planning and implementation. The LAPA Framework ensures that the process of integrating CC resilience from local-to-national planning is bottom-up, inclusive, responsive, and flexible. In the LAPA, agriculture sector is given high importance through local level planning on water uses, alternative cropping patterns, and integrating indigenous knowledge with advance technology to enhance the resilience of the vulnerable communities. Nepal's INDC also promises to undertake scientific (bio-physical as well as social sciences) approaches to understand and deal with the impacts of climate change in mountains, hills, and lowland ecosystems and landscapes. It will develop and implement adaptation strategies for climate change affected sectors particularly agriculture (GoN 2015).

National Adaptation Program of Action (Government of Bhutan 2006) of Bhutan (whole geographical area is mountainous) aims to improve the efficiency in irrigation, changes in cropping patterns, and promoting agro-forestry to reduce climate-induced soil erosion. INDC also promises to promote climate smart agricultural practices such as organic farming, climate-resilient crop varieties, conservation of plant genetic resources, surveillance of crop pests and diseases, initiating crop insurance programmes, integration of sustainable land and water management practices, and establishing cold storage facilities to contribute towards sustainable food and nutrition security (Royal Government of Bhutan 2015).

In Myanmar, there are very few agricultural policies directly related to climate change. Current agriculture sector-related policies[2] focuses on increasing productivity

[2]Agriculture polices include: National Strategy on Rural Development and Poverty Alleviation (2011), Law of Protection of Farmers' Rights and Enhancement of their Benefits (2013), Pesticide Law (1990), Farm Land Law (2012), Vacant, Fallow, Virgin Lands Management Law (2012), Climate Smart Agriculture Strategy (2016).

for food security; economic growth and rural development indirectly encompass mitigation and adaptation strategies for climate change as they relate to the improvement of the agriculture sector in Myanmar. Interventions aim to modernize agricultural practices, promote commercial farming, and liberalize investment in the sector. The recently formulated Climate Smart Agriculture Strategy (2016) focuses on adapting crop varieties and corresponding farming practices, disaster risk management, crop and income loss risk management. Myanmar's NAPA (Government of Myanmar 2012) has also aimed to increase CC resilience of rural and subsistence farmers in the Dry and Hilly Zones through leguminous crop diversification and climate-resilient varieties. However, the NAPA has not shown the innovative options of diversification in agriculture of hilly areas through cash crops (i.e. vegetables and fruits) and livestock. Myanmar's INDC is national in scope and does not specify the strategies for hilly areas. It overall reiterates the resilience of agriculture sector through implementation of climate smart agricultural practices such as leguminous crops diversification, measures in the agro-forestry sector and systematic control of soil quality and irrigation water (GoM 2015).

In Afghanistan, mainly a mountainous country, National Capacity Needs Self-Assessment (NCSA) supported by Global Environmental Facility, and NAPA (Government of Afghanistan 2009) have ranked agriculture as the second most vulnerable sector after water resources. Among CCA strategies for agriculture, the focus has been given to improve the resilience of the communities and agricultural systems. Various projects under NAPA and NCSA have set objectives including the aspects of introducing drought-resistant crop varieties and livestock breeds, establishing cold stores and agricultural cooperatives, food processing industry, maintaining product quality, crop diversifications, emergency supplies to vulnerable communities, provision of agricultural machinery and extension services, improved supply of inputs, promotion of horticulture, terracing and agro-forestry, better veterinary services, promotion of bee-keeping and poultry, rangeland management, promoting drought-resistant fodder, establishment of milk accumulation and pasteurization centres, and improved marketing facilities. However, these strategies are not specified separately for plains and mountain areas according to their topographical condition and agro-ecological potential. The most part of the geographical area of Afghanistan is mountainous, and it may be assumed that these strategies will be implemented in both plain and mountainous areas. Afghanistan's INDC emphasizes on increasing irrigated agricultural land to 3.14 M-ha, through restoration and development of Afghanistan's irrigation systems to cope with the climate-induced water variability (GoIRA 2015).

10.3.2 Status of Local Level Adaptations in Mountain Agriculture

Drawing an example of four river basins, i.e. Upper Indus (UI), Koshi, Eastern Brahmaputra (EB), and Salween Mekong (SM) from Pakistan, Nepal, India, and China, respectively, local adaptation to CC is examined. Majority of households

Table 10.3 Adaptation to climate change: evidences from mountains

Variables			Upper Indus (Pakistan) $n = 1139$	Koshi (Nepal) $n = 2310$	Eastern Brahmaputra (India) $n = 2647$	Salween Mekong (China) $n = 1987$
% HHs who observed climatic changes in last 10 years			98.2	83.0	93.0	67.6
% HH who took any actions to adapt with the climate change			66.8	32.2	33.9	52.1
% **Households**	Agricultural adaptation practices[a]	Given up planting certain crops	30.1	43.6	29.2	12.1
		Introduced new crops (%)	32.1	38.2	14.5	26.6
		Given up rearing certain livestock	24.1	22.7	3.9	3.3
		Introduced new livestock	12.0	12.4	2.5	9.6
		Changed grazing practices	15.8	17.4	0.9	1.2
		Changed farming practices	33.5	29.6	19.4	32.4
		Farmland left fallow	4.3	16.9	4.3	2.4
		Abandoned farming	2.0	8.9	2.3	1.9
		HH invested in irrigation	2.7	18.1	8.3	22.3
		Investment in irrigation	9.4	5.1	5.4	23.4

Source: ICIMOD 2014

[a]Computed among those households who took any adaptation action in response to climatic changes observed in 10 years

perceived changes in the climate over time. However, the CCA rate is not as high as it should have been. In Koshi and EB, only one-third of households took actions to adapt to the CCs (Table 10.3). There is also a vast variation across the basins as far as priority of adaptation practices is concerned. For instance, in UI and Koshi, a significant proportion of households gave up rearing those livestock which demand comparatively higher amount of water and fodder. In the same basins, considerable percentage of households changed the grazing practices of their livestock in response to fodder-stress as a result of decline in precipitation. In Koshi, around 17% households left their cultivable land fellow due to water-stress (Table 10.3). In SM, almost one-fourth households invested in irrigation to cope with the water-stress situation. Thus, it can be concluded that it is extremely important to understand the types of and variations in local CCA practices for better planning at the sub-national and national levels. Some opportunities may also arise from CC in some areas where agro-ecological potential and institutional services are better.

10.4 Status of Adjustment and Gaps

Though the HKH countries have made significant progress in designing and incorporating adaptation strategies in their national policies and programmes, there are still some gaps in these policy documents and their implementation plans. These gaps are identified and discussed below.

10.4.1 Policies, Strategies, Legal Provision, and Governance

Policies and adaptation programmes of HKH countries mainly have common policy measures for CCA for both plain and mountain agriculture. For mountains, only a limited number of policy measures have been formulated. The nature and causes of agriculture and food security in mountains is quite different from the plains and require a specific set of policy measures (Rasul et al. 2014). Food security in sparsely populated mountain areas usually does not catch much attention from researchers and development planners due to the physical isolation and higher cost of research and development activities. Mountain agriculture is more vulnerable to CC because it mainly depends on rainfall. Although the impact of the CC on agriculture and food security has become a serious concern, so far limited efforts have been made to understand how CC impact food security of the mountain communities of the HKH region, how poor households adapt to changing conditions, and what options are available to facilitate better adaptation.

Understanding the challenges to mountain farming systems and local adaptation strategies is urgently needed. There is also a need to provide new directions to formulate innovative adaptation strategies for mountain people to mitigate the impacts of changes on their livelihoods. Due to CC impacts, the mountain areas are experiencing frequent floods, accelerated deforestation; soil erosion; siltation of rivers, irrigation systems, and reservoirs; desertification; loss of biological diversity; and degradation of rangeland and pastures. Combined, these have led to a loss of soil fertility, reduced agricultural productivity, and reduced food security. Climate change is resulting in increase in temperature in the mountains, resulting in rapid glacier melting and GLOFs. What is needed is to set up appropriate mechanisms to monitor the development of glacial lakes and develop evacuation strategies in the case of GLOF in the vulnerable areas in the mountains.

10.4.2 Institutional Arrangements

In institutional arrangements for implementing adaptation strategies and measures, local institutions are not properly taken on board due to mainly two reasons. First, local institutions are not well aware of findings of scientific studies on climate change. It highlights the weaknesses of communication methods of scientists and media professionals. Second, participation of local institutions in planning and implementation of adaptation strategies is very limited. Nepal is the only example where local institutions are adequately involved in LAPAs. Moreover, adequate integration among formal government departments and organizations is also missing in almost all HKH countries. Local institutions such as local public institutions, civil society institutions and private institutions are very important in implementing adaptation strategies at the local level. They shape the livelihoods impacts of CC through a range of indispensable functions they perform in the rural context: information gathering and dissemination, resource mobilization and allocation, skills development and capacity building, providing leadership, and networking with other decision-makers and institutions. Pakistan presents a classic example of a developing country that is highly vulnerable to the adverse impacts of CC but the local institutions have done little to improve the resilience of the local communities. Adequate funds and mandates are not granted to provincial and local institutions to implement strategies for adaptation to climate change. Strong traditions of local governance, communal resource tenure, and local ecological knowledge are a few of the traits that decision-makers and implementers have not appropriately used (Hussain et al. 2016).

There are some common issues which are found with the operations of all type of local institutions in HKH region. Among those, the most important are inadequate participation of women, exclusion of poor segments and lower castes in decision-making, and ethnic divisions in resource management and utilizations. Within poor communities in the mountain areas, women and the poor segments of the society tend to be particularly vulnerable to environmental degradation and natural disasters. When developing and implementing adaptation strategies at the local level, it is critical to recognize and respect the greater vulnerability of women and the poor to the impacts of climate as well as the difference in the way they are affected in comparison to other groups. It is mostly the influential community members and farmers who succeed in obtaining support and services from local government and non-government institutions. There seems to be no equity in local adaptation strategies. Thus, give rise to a very important question: "How can the poor and marginalized mountain farmers be included in the adaptation process?"

10.4.3 Integration and Mainstreaming

Despite continuous efforts from several stakeholders at different levels, the vulnerable mountain communities are still unable to adapt to CC effectively due to flaws in integration across the institutions, sectors, and policies. Some countries, for example Pakistan, have shown notable integration of adaptation strategies in their policies of CC, and Agriculture and Food Security for achieving sustainable food security in the mountain areas. The countries with NAPAs are also needed to incorporate and integrate policy steps in action plans of other relevant ministries and sectors. Without proper integration, it is not possible to achieve long-term adaptation goals.

10.4.4 Finance

Over the years, it has been observed that CCA could not be adequately incorporated in the development goals of the HKH countries. Climate change is uncertain and its impacts are long term. However, development goals in the region are mostly set based on the obvious facts having short-term consequences. Additionally, CC demands global actions with trans-boundary cooperation, but national adaptation measures and strategies mostly focus on the parameters within national context. Due to such factors, CCA do not attract adequate internal funds from national and sub-national sources, resulting in low allocation of funds for adaptation in the vulnerable zones such as mountains. The HKH countries are mainly dependent on external funding sources including several funds established under UNFCCC such as Least Developed Country Fund (LDCF), Climate Change Special Fund (SCCF). Higher dependency on external financial sources is leading to delayed or weak actions in implementing CCA strategies. For example, recently developed CC Policy of Pakistan and Nepal mainly target the external funding sources. It has resulted in delays of its implementation phase.

10.4.5 Implementation Capacity

In the HKH countries, capacity of institutions to implement the policies and programmes is inadequate due to lack of integration across sectors, shortages of funds, lack of skills and technical knowledge among the professionals and implementing staff, and inadequate supportive infrastructure. In some countries such as China and India, the institutions have shown better implementation capacity. However, other countries need to improve the capacity of institutions for better implementation particularly in the vulnerable zones including mountains.

10.4.6 Technology

The key challenge faced by the policy makers and development practitioners in the HKH region is how to increase agricultural productivity and reduce poverty without degradation of the natural resources on which mountain farmers largely depend. The research and development interventions in the HKH for agriculture have not been very effective because these are mainly focused to agricultural technologies and services for plains (Rasul and Karki 2007). To achieve this goal, it is necessary to understand the characteristics of the mountain agriculture, its specificities, socio-cultural and policy dimensions, and niche potentials for harnessing comparative advantages to support livelihoods, and protect environment. It has been found that mountain areas in the HKH region have high agro-ecological potential for cultivation of fruits, nuts, vegetables, tea, tobacco, and cereals such as native varieties of rice, buckwheat, barley, and sorghum. Technologies for efficient management of water, soil nutrients and pests, and integration of livestock may enhance flexibility of mountain agriculture for better adaptation. To implement these technologies as a package, the concept of climate smart agriculture is emerging in mountain areas. Nepal is the first country where the concept of climate smart agriculture is implemented successfully in mountains. It is equally important to analyse the critical issues, gaps in policy, programmes, implementation strategy, and demand-driven technologies and services for achieving sustainable outcomes. Similar to plain areas, there is a need to promote technology and innovations in the areas of mechanization, photosynthesis, biological nitrogen fixation, bio-technology, prevention of diseases and pests, stress resistance, precision agriculture, and post-harvest technologies in mountain areas.

Climate Smart Villages in mountains of Nepal

In 2014, the International Centre for Integrated Mountain Development (ICIMOD) and Centre for Environmental and Agricultural Policy Research, Extension and Development (CEAPRED) implemented the Climate Smart Village program in five villages, i.e. Nayagaon-Deopur, Patalekhet, Kalchebesi, Naubise and Dhaitar, of Kavrepalanchok district in Nepal. This is the first effort to establish climate smart villages in the mountain areas. In the selected villages, climate smart activities such as water smartness/management, appropriate cropping patterns, energy smartness, nutrient management, ICT services and farmers' linkages to experts, and social networking/cohesion were introduced.

Before introducing the climate smart interventions to villages, a small field survey was conducted to study the existing water management practices, cropping patterns, energy sources, nutrient and soil management, ICT services and functionality of farmers' groups. Out of around 500 households in the selected villages, 63 households (13 % of total households) were randomly surveyed in the villages. Level of adoption of climate smart practices in the selected villages before intervention is presented in the table below.

Most of the farmers were not practicing climate smart activities on their farms. In addition, women's participation was also limited in household and farm decisions. ICIMOD and CEAPRED have strengthened the capacity of the villagers through trainings, demonstrations, technical input, ICT, and market linkages to increase their adoption rate of climate smart activities.

Recent, field observations reveal that farmers are adopting climate smart activities and they are satisfied with the performance and results of those activities. Moreover, they have succeeded to maintain their income levels with low cost climate smart activities. After the successful implementation of this program, ICIMOD is supporting the government of Nepal to out-scale climate smart activities in other mountain areas within Nepal. Recently, this concept has also been implemented in hilly areas of Pakistan by ICIMOD and Pakistan Agricultural Research Council (PARC).

Level of adoption of climate smart practices before interventions				
Climate smart practices	**Distribution of households (%) by level of use**			
	Frequently used	**Moderately Used**	**Rarely used**	**Never used**
Zero tillage	0	0	0	100
Least tillage	0	5	3	92
Organic manure/vermicomposting	0	0	3	97
Green manure	100	0	0	0
Use of IPM technology	0	13	19	68
Use of slurry & urine	14	14	30	41
Use of bio-pesticides	0	3	10	87
Use of bio-fertilizer	0	2	8	90
Crop rotation (Nitrogen fixing crops)	11	38	27	24
Intercropping./Relay cropping	2	14	29	56
Line (horizontal/vertical/diagonal) and contour	30	14	30	25
Retained straw residue instead of burning	30	5	13	52
Other mulching techniques	0	14	3	83
Rain water harvesting	2	0	0	98
Drip/pitcher irrigation	0	0	0	100
Digging ridges rather than flood irrigation	30	10	30	30
Water storage tanks	2	2	8	89
Water storage in small ponds at farm	2	2	0	97
Terracing and bonding	40	5	11	44

10.4.7 Decision-Making Tool

In the HKH countries, reports produced by national institutions based on national and sub-national level official data sources provide input for decision-making process to plan the adaptation strategies. Some regional or international organizations also conduct policy studies to support and provide input to policies and programmes of the governments. Sometimes national data sources and official reports do not reveal reliable statistics because these reports follow the common framework for both plain and mountain areas to assess the vulnerability of the people and systems. Mountain areas have different climatic and socioeconomic characteristics and agro-ecological potential, and need special attention. What is therefore needed is to formulate different framework and tools to assess the vulnerability in the mountains which may provide robust input to policy making process for CCA in agriculture, conducive to food security.

10.4.8 Awareness and Access to Information

Despite several constraints, the mountain areas have high potential in natural resources such as water for irrigation and hydro-power generation and high value horticulture. Due to limited resources and poor institutional capacity, these resources are not being exploited properly for sustainable food security of the people. There is a need to create proper awareness of mountain farmers on the climate projections and arising opportunities. Capitalizing on the arising opportunities may turn into the most appropriate adaptation strategies. In Pakistan, China, India, and Nepal, for instance, changes in temperature and precipitation patterns have added to the agro-ecological potential of several valleys which is suitable for high value horticultural crops, but farmers are still practising conventional agriculture. On the other hand, in some mountain areas which are facing water-stress water intensive crops (i.e. rice and pulses) are still being promoted, e.g. high mountain pockets of Nepal. A proper research-based guideline is needed to promote area specific agriculture and livelihood strategies in mountain areas with different levels of agro-ecological potential and impacts of CC in the HKH countries.

The communication of climate science to policy makers is also very important. It entails close connections between science and various other realms of society, such as local customs, politics, and the economy (Ivanova et al. 2013). The debate on the relationship of climate science, policy makers, and vulnerable farming community was started particularly when CC emerged into the global development agenda. Media coverage has a direct effect on farmers' opinion on particular issues. Using the media, awareness among the mountain farmers may be created about the CC impacts and arising opportunities.

10.5 Readiness for Enhanced Adaptation Action

The concerned ministries and departments in the HKH countries have their own framework of implementation of adaptation programmes and projects. Gaps mentioned in the previous section need to be addressed and some additional measures to be taken for enhanced agricultural adaptation practices in the mountain areas to achieve sustainable food security. Additional measures for enhanced adaptation are suggested below.

Policies, strategies, legal provision, and governance For enhanced adaptation action, HKH countries need to focus on land-use planning for mountain regions. A policy or programme cannot benefit the vulnerable communities until unless it follows a framework with mountain-specific strategies formulated on the basis of agro-ecological potential and level of CC impacts. China has somewhat started formulating specific strategies and has started conservation practices in those mountain areas which have very limited potential for agricultural productivity. The areas with higher potential and better market services are becoming centres of agricultural commercialization.

Some mountain areas with higher agro-ecological potential could not get proper attention from the adaptation programmes, and majority of farmers, without knowing the potential, are still practising conventional agriculture. As a consequence, youth in such areas has lost their interest in agriculture. For instance, in some mountain areas of Pakistan such as Chitral (KPK) and Hunza-Nagar (GB), which have the highest agro-ecological potential for fruit, vegetables, and other crops, almost 70% of farming households face frequent labour shortages during the sowing and harvesting months due to outmigration of active household members, particularly youth (Rasul et al. 2014).

There is need to understand the adaptation co-benefits across the sectors. For example, hydro-power and solar energy are alternative energy sources for mountain agriculture. In dry mountain ranges with good agro-ecological potential (i.e. areas of Gilgit-Baltistan in Pakistan), these energy sources will play very crucial role to pump groundwater and fetch spring/river water to agriculture terraces on higher altitude. If energy sector stores and utilizes the excessive water during monsoon period and erratic rainfalls in other seasons, it will not only ensure energy security in the particularly country but also provide additional energy supply to agriculture and other sectors. National policies and programmes should focus on the co-benefits across the sectors for enhanced adaptation and food security in mountains. Mountains of HKH countries have huge potential for hydro-power but this potential is not utilized adequately. For instance, Nepal has hydro-power potential of 43,000 MW but the installed capacity is of only 600 MW (HIDCL Nepal 2015). Likewise, Pakistan has hydro-power potential of nearly 60,000 MW but only 11% (6720 MW) of the total potential has been exploited (Rasul et al. 2014).

Trans-boundary cooperation among the HKH countries and realization of co-benefits of adaptation are also very important factors for achieving sustainable and enhanced adaptation to CC in the region. Initiative of regional countries on information sharing, flood warning systems, management of water resources, and other issues will enhance the resilience of the region as a whole against climate change impacts. For enhanced adaptation actions in mountain agriculture, it is also inevitable to protect and promote local food systems with higher agro-diversity in selected areas. Despite having higher nutritional value, some species such as finger millet, buckwheat, high altitude rice, naked barley, rice bean, horse crop, and so many others are disappearing from mountains. To maintain the diversity in diet and reduce the dependency on commercial crops, future strategies should also focus on local food systems.

Institutional arrangements Currently, the mountain people and local institutions are not well prepared to cope with climate change impacts. Adequate funds and mandates are not granted to provincial and local institutions to implement strategies for adaptation to climate change (Hussain et al. 2016). However, Nepal has made significant progress in this regard through the LAPAs which have brought local informal institutions into adaptation planning process. In LAPA, agriculture sector has given high importance through local level planning on water uses, alternative cropping patterns, and integrating indigenous knowledge with advance technology to enhance the resilience of the vulnerable communities. Involvement of local institutions also provide opportunities to women to participate in the local planning for agricultural adaptation strategies.

Capacity of local institutions should be enhanced to help the local communities not only in adaptation but also to respond the climate change-induced disasters in the mountain areas. During recent floods in Myanmar and Pakistan (in Chitral district), the people faced a lot of challenges because local institutions were not capable of helping the people for mitigating at least their economic and human losses.

Integration and mainstreaming Incorporation of CCA strategies in programmes and policy of other relevant sectors is also important. In Pakistan, for instance, mountain-specific CCA strategies in National Climate Change Policy are also incorporated in the National Agriculture and Food Security Policy. Stand-alone adaptation actions may not achieve the objectives of enhanced adaptation because integrated approach by sectors such as climate change, agriculture, water, energy, and food security are needed for better outcomes. Moreover, during the implementation phase of CCA strategies, aspects of horizontal[3] and vertical[4] integration also need to be ensured which would improve monitoring of the implemented activities. In the HKH countries, integration across sectors is very limited, and the most affected sectors by climate change such as water, agriculture/food, and energy are working almost independently (Rasul and Sharma 2016).

Finance In terms of financial aspects, most HKH countries are heavily dependent on external financial resources. It is important for national and sub-national governments to allocate adequate funds for CCA particularly for vulnerable mountain zones. In this regard, proper awareness needs to be created among the policy makers that CCA in mountains is equally important like other sectors such as infrastructure, health, and education because climate-induced changes in the mountains are inducing equally severe impacts in the downstream areas in the form of floods or shortage of water in irrigation systems (Hussain and Routray 2012). The government of

[3] Coordination and synergies across the sectors, ministries. and other relevant organizations.

[4] Coordination between the different levels (implementing department/ministry, coordinators, professional/technical staff, regional staff, field officers, etc.) within the same organization.

Khyber-Pakhtunkhwa (KP) province of Pakistan has set a good trend to give an equal importance to climate and environment like other sectors of economy. They allocated 15 billion Pak rupees for "Billion Tree Tsunami" project with the target of planting 1.2 billion trees in KP by 2018. They also allocated sizeable funds for renewable energy production, e.g. micro-hydro power.

Implementation capacity In China, the institutions have shown better implementation capacity. However, other countries need to improve the capacity of institutions for better implementation of CCA strategies in the vulnerable zones such as mountains.

Technology Currently, only a few countries like China, Pakistan, and India have adequate technology uptake for enhanced adaptation practices; however, they still need to strengthen their efforts in the mountain areas. Overall, all HKH countries need to promote technology and innovations in the areas of solar energy, mechanization, photosynthesis, biological nitrogen fixation, bio-technology, prevention of diseases and pests, stress resistance, precision agriculture, post-harvest technology, and water conservation in the mountain areas. For example, in Mithinkot Village Development Committee of district Kavrepalanchok in Nepal, farmers were facing water shortages in agriculture. With the support of the Centre for Environmental and Agricultural Policy, Research, Extension and Development (CEAPRED) and District Agricultural Development Office, farmers started growing fresh vegetables in 2008 after starting rainwater harvesting and constructing small irrigation ponds. Later on, they also started producing hybrid tomato seed of Srijana variety. It led to a significant improvement in their food security and livelihoods. In Pothwar region of Pakistan, the government institutions promoted solar powered irrigation pumps, water conservation ponds, and recently climate smart cropping patterns. Several farmers adopted these technologies which resulted in improvement of their income and food security.

Decision-making tool The concerned ministries and departments are relying on their own vulnerability assessment reports or other official data sources in the countries. In the HKH countries, it has been observed that common frameworks are being used for vulnerability assessment in both plain and mountain areas. There are chances of biased estimates because mountains have different socioeconomic and climatic factors compared to plains. Only good quality reports with separate vulnerability assessment framework for mountains provide robust input in decision-making tools and mechanisms.

Awareness and access to information Mountain farmers are not well aware of the climate projections, its impacts, and arising opportunities from changing agro-ecological potential (Rasul et al. 2014). Despite the severe impacts of CC on the conventional crops, i.e. cereal, they have been cultivating the same crops over the years. It is important to identify the arising opportunities in the form of high value cash crops such as fruits, vegetables, and nuts in the changing agro-ecological potential due to temperature rise and changes in precipitation.

Based on authors' judgement, level of readiness for enhanced adaptation action in mountain areas of HKH countries has been guessed for all eight indicators

(Table 10.4). Average score for mountains in the last column shows that the HKH region, overall, is not adequately ready for enhanced adaptation action. There is a need to take solid steps to improve all eight indicators of readiness.

10.6 Attaining Adaptation Vision

Rising impacts of CC have further added to food insecurity and poverty in both upstream and downstream areas of HKH region due to temperature changes and frequent floods, severity and fluctuation of seasons, and unusual changes in precipitation. Changes occurring in mountain areas are also inducing severe impacts on downstream areas through increasing pressure on natural resources and production systems, and competition for food, accommodation, and income opportunities.

Realizing the importance of adaptation to CC, the countries of the HKH region have started taking concrete steps towards formulating and implementing adaptation strategies. Countries in the HKH have submitted their respective NAPAs and/or formulated their own policies and programmes to cope with the CC impacts. However, there is a need to further strengthen the CCA strategies in the mountain areas to enhance the resilience of agriculture systems for achieving sustainable food security. Majority of people living in the mountain areas of HKH region, irrespective of countries, are depending on agriculture for their livelihoods and food security. For the governments of HKH countries, it is extremely important to bring mountain-specific strategies in their policy documents because the nature of agriculture and factors of food security in mountains are very different from plains. Essentially, mountain areas can be divided into four types depending on whether they have high or low agro-ecological potential and good or poor access to markets, information, technology, and institutional services. Different development strategies are needed for each of these different types of area (Table 10.5). Strategies for areas with high potential and good access to market and services should focus on exploiting the existing potential as much as possible through land-use intensification, efficient water use, crop diversification, commercial dairy, and growing of cash crops that offer higher incomes. Private investment in production and post-harvest facilities should be encouraged. In areas with high potential but poor access to markets and services, the focus should be on removing the marketing constraints and developing infrastructure and institutional support so that the existing potential can be exploited optimally and sustainably. In areas with low potential but good access to markets and services, strategies should focus on improving the use of existing facilities to provide economic incentives and appropriate regulations to promote the sustainable use and management of resources. Finally, in areas with low potential and poor access to markets and services, strategies should focus on the subsistence use of resources to reduce the dependence on local resources and ensure food security.

Regarding institutional arrangements for implementing adaptation strategies and measures, local institutions have not been properly taken on board. Local institutions such as local public institutions, civil society institutions, and private institutions are very important in implementation of adaptation strategies at local level. In

Table 10.4 Readiness for enhanced adaptation action in mountains

Indicators	Level of readiness in mountain areas (Ready = 4; Moderately ready = 3; Less ready = 2; Not ready = 1)								
	China	Pakistan	India	Bangladesh	Nepal	Bhutan	Myanmar	Afghanistan	Average score for mountains
Policies, strategies, legal provision, and governance	3	4	2	1	3	4	2	2	2.6
Institutional arrangements	1	1	1	1	3	1	1	1	1.3
Integration and mainstreaming	3	3	2	2	2	2	1	1	2.0
Finance	2	2	1	1	1	1	1	1	1.3
Implementation capacity	3	2	3	2	2	2	2	2	2.3
Technology	4	3	3	2	3	3	2	1	2.6
Decision-making tool	2	2	2	2	2	2	1	1	1.8
Awareness and access to information	2	2	2	2	3	3	1	1	2.0

Source: Authors' judgement

Table 10.5 Area-specific adaptation approaches in agriculture for mountain food security

	Access to markets, information, technology, and institutional services	
Agro-ecological potential	Good	Poor
High	• Enhance support for high value cash crops, e.g. fruit and vegetable cultivation • Establish fruit processing and storage facilities • Promote horticulture, commercial dairy, and intensive food crop production • Encourage private investment in irrigation, land management, and the agro-processing industry through institutional support	• Improve marketing facilities for fresh fruit, nuts, dried fruit, vegetables, and livestock products • Improve transportation facilities, ICT access, and others • Promote subsistence food crops • Promote technology and develop infrastructure to enable utilization of the mountains' high agro-ecological potential • Promote livestock, livestock products, and by-products
Low	• Promote conservation technologies/ climate smart practices that enhance agricultural potential and utilize local niches, and provide incentives for conservation • Encourage agro-forestry, tree farming for timber, and NTFPs and medicinal plants • Protection and promotion of local food systems with higher agro-diversity	• Provide incentives for conservation, maintain biodiversity and sustainable use of resources, and develop mechanisms for payments for ecosystem services • Promote subsistence agriculture with zero-tillage, mixed cropping, and livestock production

Source: Adopted from Rasul and Hussain (2015) and modified

addition to incorporating adaptation actions in national policies, it is also important to further strengthen synergies and integration across the sectors, ministries, and departments for collective efforts towards adaptation to CC. Higher dependence on external financial sources is leading to delayed or weak actions in implementing CCA strategies. Internal allocation of adequate funds to CCA will enhance the strategies and their outcomes. In some countries such as China and India, the institutions have shown better implementation capacity. However, other countries need to improve the capacity of institutions for better implementation. Institutional capacity may be enhanced through better integration across sectors, adequacy of funds, improvement of skills and technical knowledge among the professionals and implementing staff, and adequate supportive infrastructure.

Diversification of adaptation practices is an important step towards enhancing the resilience of the vulnerable communities to counter severe impacts of CC such as drought, floods, low productivity, and food insecurity. For diversification of adaptation, adoption of technology and innovations is an important step. In the mountain areas, the governments have to promote technology and innovations in the areas of mechanization, photosynthesis, biological nitrogen fixation, bio-technology, prevention of diseases and pests, stress resistance, precision agriculture, post-harvest technology, and water conservation and irrigation. In areas with higher degradation of natural resources, climate smart technologies may be promoted. For decision-making on adaptation strategies and their implementations, the governments need to use reliable data sources as an input for adaptation planning for mountains. Reliable data means the results of the vulnerability assessment tool which is specifically designed for mountain areas based on their socio-economic and climatic factors and agro-ecological potential.

For better understanding of mountain farmers, there is a need to communicate climate science in a simple and understandable way. There is also a need to create awareness among the farmers about the rising opportunities in the form of high value crops, i.e. fruits, vegetables, and nuts, due to changing climatic conditions.

References

DAWN (2015) Chitral floods: why melting glaciers may not be the cause. http://www.dawn.com/news/1197805. Accessed 10 Aug 2015

FAO (2013) FAO statistical year book 2013. Food and Agriculture Organization of the United Nations, Rome

FSA (2009) Food insecurity in Pakistan, food security analysis 2009. Sustainable Development Policy Institute-Swiss Agency for Development and Cooperation-World Food Programme, Pakistan

Giribabu M (2013) Food and Nutritional Security in North East India: some contemporary issues. Int J Dev Res 3(5):1–8

Government of Afghanistan (2009) National Capacity Needs Self-Assessment for Global Environmental Management (NCSA) and National Adaptation Programme of Action for Climate Change (NAPA), Afghanistan

Government of Bhutan (2006) National adaptation program of action. National Environment Commission, Royal Government of Bhutan

Government of India (2008) National action plan on climate change. Prime Minister's Council on Climate Change, India

Government of Islamic Republic of Afghanistan (GoIRA) (2015) Intended Nationally Determined Contribution, Climate Change Division National Environmental Protection Agency of Afghanistan

Government of Myanmar (2012) Myanmar's National Adaptation Programme of Action (NAPA) to climate change. National Environmental Conservation Committee, Ministry of Environmental Conservation and Forestry, Myanmar

Government of Myanmar (GoM) (2015) Myanmar's Intended Nationally Determined Contribution (INDC). Ministry of Environmental Conservation, the Government of the Republic of the Union of Myanmar

Government of Nepal (2011) Climate change policy. Government of Nepal

Government of Nepal (GoN) (2015) Intended Nationally Determined Contributions (INDC). Ministry of Population and Environment, Government of Nepal

Government of Pakistan (2012) National climate change policy. Ministry of Climate Change, Islamabad, Pakistan

Government of Pakistan (GoP) (2015a) Draft agriculture and food security policy of Pakistan. Ministry of National Food Security and Research, Islamabad, Pakistan

Government of Pakistan (GoP) (2015b) Pakistan—intended nationally determined contributions. Ministry of Climate Change, Islamabad, Pakistan

Government of People's Republic of China (2007) China's national climate change programme. National Development and Reform Commission, China

Government of People's Republic of China (GoPRC) (2015) China's intended nationally determined contributions: enhanced actions on climate change. Department of Climate Change, National Development and Reforms Commission, Beijing, People's Republic of China

Government of the People's Republic of Bangladesh (2009) Bangladesh climate change strategy and action plan 2009. Ministry of Environment and Forests, Dhaka, Bangladesh

Government of the People's Republic of Bangladesh (2015) Intended nationally determined contributions (INDC). Ministry of Environment and Forests (MOEF), Government of the People's Republic of Bangladesh, Dhaka, Bangladesh

HIDCL Nepal (2015) Nepal hydropower overview. Hydroelectricity Investment & Development Company Limited. http://www.hidcl.org.np/nepal-hydropower.php. Accessed 11 Aug 2015

Hussain A (2010) Achieving food security in Pakistan: assessment of current situation and policy issues at national and regional levels. Lambert Academic Publishing, Germany

Hussain A, Routray JK (2012) Status and factors of food security in Pakistan. Int J Dev Issues 11(2):164–185

Hussain A, Agarwal NK, Leikanger I (2016) Action for adaptation: bringing climate change science to policy makers—a synthesis report of a conference held in Islamabad on 23–25 July 2015. Food Sec 8:285–289

ICIMOD (2008) Food security in the Hindu Kush Himalayan region. A position paper. International Centre for Integrated Mountain Development (ICIMOD), Nepal

ICIMOD (2014) Results of vulnerability and adaptive capacity assessment (VACA). International Centre for Integrated Mountain Development (ICIMOD), Kathmandu, Nepal

ICIMOD (2015) Results of vulnerability and adaptive capacity assessment (VACA). International Centre for Integrated Mountain Development (ICIMOD), Kathmandu, Nepal

IPCC (2007) Agriculture. In: Metz B, Davidson OR, Bosch PR, Dave R, Meyer LA (eds) Climate change 2007: mitigation. Contribution of working group III to the fourth assessment report of the intergovernmental panel on climate change. Cambridge University Press, Cambridge and New York

IPCC (2014). Asia. In: Barros VR, Field CB, Dokken DJ, Mastrandrea MD, Mach KJ, Bilir TE, Chatterjee M, Ebi KL, Estrada YO, Genova RC, Girma B, Kissel ES, Levy AN, MacCracken S, Mastrandrea PR, White LL (eds) Climate change 2014: impacts, adaptation, and vulnerability. Part B: regional aspects. Contribution of working group II (authors: Hijioka Y, Lin E, Pereira JJ, Corlett RT, Cui X, Insarov GE, Lasco RD, Lindgren E, Surjan A) to the fifth assessment report of the intergovernmental panel on climate change. Cambridge University Press, Cambridge and New York, pp 1327–1370

Ivanova A, Schäfer MS, Schlichting I, Schmidt A (2013) Is there a medialization of climate science? Results from a survey of German climate scientists. Sci Commun 35(5):626–653

Ministry of Environment (2010) National adaptation programme of action to climate change. Kathmandu, Nepal

MOHP Nepal (2012) Nepal demographic and health survey 2011. Ministry of Health and Population, New ERA, and ICF International, Calverton, Maryland, Kathmandu, Nepal

Rasul G, Hussain A (2015) Sustainable food security in the mountains of Pakistan: towards a policy framework. Ecol Food Nutr 54(6):625–643. Article in Press

Rasul G, Karki M (2007) Pro-poor policy agenda for sustainable agriculture development in the Hindu Kush-Himalayan region: talking points 2/07. ICOMOD, Kathmandu

Rasul G, Sharma B (2016) The nexus approach to water–energy–food security: an option for adaptation to climate change. Clim Pol 16(6):682–702. https://doi.org/10.1080/14693062.2015.1029865

Rasul G, Hussain A, Khan MA, Ahmad F, Jasra AW (2014) Towards a framework for achieving food security in the mountains of Pakistan, ICIMOD Working Paper 2014/5. ICIMOD, Kathmandu

Rasul G, Saboor A, Tiwari P, Hussain A, Ghosh N, Chettri G et al. (2018) Chapter 9- food and nutrition security in the HKH. In: Hindu Kush Himalayan Monitoring and Assessment Programme (HIMAP) (forthcoming). Kathmandu: International Centre for Integrated Mountain Development (ICIMOD).

Royal Government of Bhutan (2015) Intended nationally determined contribution. National Environment Commission, Ministers of Agriculture and Forests, Thimpu, Bhutan

Shafiq M, Kakar MA (2006) Effects of drought on livestock sector in Balochistan province of Pakistan. Int J Agric Biol 8(6)

Thethirdpole.net (2015) Glacier floods cause humanitarian crisis in Chitral. http://www.thethirdpole.net/glacier-floods-cause-humanitarian-crisis-in-chitral/. Accessed 10 Aug 2015

Tiwari P (2000) Land-use changes in Himalaya and their impact on the plains ecosystem: need for sustainable land use. Land Use Policy 17(2):101–111

Tiwari PC, Joshi B (2012) Natural and socio-economic factors affecting food security in the Himalayas. Food Sec 4(2):195–207. https://doi.org/10.1007/s12571-012-0178-z

World Bank (2015) Agriculture data of the World Bank. http://data.worldbank.org/indicator/NV.AGR.TOTL.ZS. Accessed 21 June 2015

Chapter 11
Enhanced Actions on Adaptation

Mozaharul Alam

Abstract Asia and the Pacific region, home of large world population, has made progress in both social and economic development while quality of environmental resources went down. In addition to degradation of environmental resources, impacts of climate change are becoming more and more visible. Countries in the region are taking steps and started adjusting policies and building institutional capacity but the pathway for attaining the adaptation vision, goals, and objectives still requires overcoming several bumps and humps along the way. The constraints facing by the region to enhance its adaptation efforts consisting of a complex web of interconnected factors which influence each other. While a clear timebound adaptation vision is necessary to develop by most of the countries in the region, making mainstreaming or integration of climate change adaptation is business for all actors is equally important. Investing on institutional capacity building for adaptation planning, implementation, and monitoring results is necessary too. A country would also need to take advantages of interlinkages and multiple benefits of measures of climate change adaptation, development, and disaster risk reductions in achieving its sustainable development goals. This chapter highlighted several areas that countries in the region need to focus on to advance its adaptation actions.

Keywords Adaptation actions · Finance · Integration · Institutional capacity mainstreaming · Technologies

11.1 Introduction

Asia and the Pacific region has experienced rapid economic development and has alleviated poverty during the last two decades. In contrary, degradation of environmental resources caused by unrestrained growth becoming severe which is undermining development. Observed and anticipated adverse impacts of climate change are undermining sustainability of development efforts and its achievement in long

M. Alam (✉)
United Nations Environment Programme, Bangkok, Thailand
e-mail: alam31@un.org

M. Alam et al. (eds.), *Status of Climate Change Adaptation in Asia and the Pacific*, Springer Climate, https://doi.org/10.1007/978-3-319-99347-8_11

run. Poor and developing countries, particularly the Least Developed Countries (LDCs) and Small Island Developing States (SIDS) will be among those most adversely affected and least able to cope with the anticipated climate risk to their social, economic, and natural systems. For examples, sea level rises could submerge the lowest parts of the Mekong delta, and millions of people will be displaced with densely populated and low-lying countries like many Small Island Developing States (SIDS) (IPCC 2014).

Countries in the region are not sitting idle and have made progress in several areas to deal with adverse impacts of climate change. However, enhanced actions on adaptation to match with huge needs and prevailing gaps, countries in the region need to enhance its efforts on building national level institutional capacity. In addition, a clear and timebound adaptation goal along with national and sectoral targets is very important for monitoring and reporting progress of adaptation and revisiting its efforts in building long-term resilience. While a country will be giving more emphasis on adaptation interventions focusing on key vulnerable sectors and its people for their urgent and immediate needs, it is also important not to lose sights on medium- and long-term adaptation needs as well as emergence of new issues arising from climate change.

It is also important to recognize that adjustment requires time and involved changes, modification, and improvement in many areas. It requires (a) understanding of impacts and vulnerability for adaptation planning; (b) access to finance and technologies, (c) access to relevant information, knowledge, technical skill and knowhow for designing and implementing adaptation interventions, and (d) monitoring, evaluating, reporting, and revising impacts or results. It is also important to understand policy and institutional landscape as well as engagement of relevant actors while designing and implementing adaptation interventions. Adaptation interventions also need to be specific to actors and beneficiaries at different levels as impacts are multifaceted and cut across social, economic, and environmental dimensions.

11.2 Essentials for Enhanced Actions

11.2.1 Place Clear Timebound Goals and Targets

Without clear timebound adaptation goals and objectives, assessment of status of adaptation and measuring progress towards stated goals and objectives would be a difficult exercise including comprehending adequacy of progress. While drivers of vulnerabilities vary by sub-region and country, the overarching adaptation goals are committed to protect people, development, and ecology from adverse impacts of climate change (Sect. 3.3.1). Being an agrarian society and significant contribution of agriculture to economy and jobs for rural communities, the common vision for agriculture is to enhance resilience of the food production system through (a) enhancement of crop yields to sustain the population demand; (b) economic stability, and (c) reduction in vulnerability of farming communities. It is worth noting that adaptation strategies in agriculture are closely associated to water resource management as water resources are also highly susceptible to climate change.

Though, most countries in Asia and the Pacific region have overarching goals and objectives, they do not have clear timeframe to achieve those goals and objectives including absence of sector-specific targets. However, opportunities are not lost. The Nationally Determined Contribution (NDC), a timebound and target oriented national document, provides an excellent opportunity for each country to come up with a timebound plan with specific targets towards achieving overarching national, and sectoral adaptation goals and objectives. This also provides opportunity to revisit progress in every 5 years and enhance ambition.

11.2.2 Makes Mainstreaming Business for All

Both climate change adaptation measures and the concept of climate resilient development address reduction of vulnerability, enhance adaptive capacity, and support resilient development of vulnerable sectors and societies. Therefore, mainstreaming and integration of climate change adaptation into development planning and budgeting system is very important. "Business as usual" development will not be able to address future needs and challenges arising from climate change, and therefore it requires adjustment and even sometime transformative change. Therefore, a proactive approach is necessary to mainstream adaptation to climate change into national and sectoral development strategies, programme, and projects. Further, an appropriate mechanism is necessary to engage wider participation of all stakeholders including local institutions, communities, and to consider local circumstances while formulating adaptation plan, and implementing adaptation interventions. Engagement of all stakeholders will support closing gaps between national policies and strategies on adaptation and community needs at local level.

Special efforts must be taken to empower local institutions and enable them to effectively take part in adaptation planning and implementation of adaptation measures at local level. It is vital to ensure that mainstreaming is not only happening into the national programmes and policies, but vertical integration of climate risks and responses has also been considered into the local development plans. Countries in the Pacific and with low-lying coastal areas should not carry out stand-alone climate proofing of coastal infrastructure projects. These infrastructure projects need to be part of a comprehensive coastal management plan within an integrated coastal management policy and backed up by an appropriate and effective legislation.

All development projects need to consider the potential impacts of climate change on resources on which development projects are relying on. Potential entry points to consider potential impacts could be the Environment Impacts Assessment (EIA) guidelines, national and sectoral project development guidelines and pro-forma, and strategies and guidelines of bilateral development partners. Incorporation of climate risks—current and future—into these strategies and guidelines will advance mainstreaming and integration. It is also necessary to build capacities of the designated bodies to oversee mainstreaming and integration of climate change adaptation across different levels, i.e. national and sectoral development plan, sub-national, township, villages, and among different ministries and agencies.

11.2.3 Target the Most Vulnerable

Protecting people, ensuring food security, and building resilience of infrastructure are key long-term adaptation goals and objectives of almost all countries in Asia and the Pacific region. These include protecting people and avoiding impacts on socio-economic development from water-related extreme events and hazards; protect and manage the coasts including human settlement and infrastructure from tropical cyclone, storm surges, inundation, etc. While designing and implementing adaptation interventions at local level, it is very important to engage local government to promote a more effective and efficient way of implementing projects that benefit communities. Particularly, for any food security project, farmers and farmer associations should be involved as much as possible and conduct location specific on firm trial than the research station. Further, for enhancing resilience of the vulnerable communities to deal with adverse impacts of climate change, diversification of adaptation practices is an important step to consider.

It is entirely logical and justifiable to focus adaptation interventions on key vulnerable sectors and people due to limited availability of fund including challenges to access fund. However, a country needs to keep eyes on and revisit its vulnerability, appraising adaptation needs, and not to lose eyes on likely emerging issues due to dynamic nature of changes in the climate system and development.

11.2.4 Invest in Institutional Capacity Building

Mainstreaming climate change adaptation into national and sectoral policies, strategies, programme, and plans need to be supported by (a) strengthening interagency coordination and communication to support horizontal and vertical aspects of mainstreaming; (b) strengthening technical capacity for adaptation planning, and (c) empowering institutes for undertaking an indicator-based monitoring and reporting system.

Create an enabling policy and organizational framework, fostering dialogue among institutions dealing with the development sectors which are at risk to climate change such as agriculture, water, health, infrastructure, and energy and establish a mechanism for regular information sharing and knowledge exchange both horizontally (national level) and vertically (national-sub-regional/local) are essential for strengthening interagency coordination and nurturing collaboration. This horizontal and vertical interaction among institutions will create rooms for multiple actors to play their respective roles in a coordinated manner as well as avoid fragmented efforts, sectoral silos, and unimodal actions.

Strengthening institutional coordination and communication need to be supported further by strengthening coordination among technical working groups, and multi-sectoral steering committees, many of which are merely at the initial step. There is also a need to review mandates, roles, and responsibilities of each institution involved these committees to avoid overlap of functions and smoothening its operation. A designated entity with the authority to conduct overall coordination will also ease its operation. For proactive engagement in integrating climate change adaptation into development, skills enhancement programme for policy makers and legislators is necessary. In addition to national level capacity development programme, regional capacity building programme on climate change on targeted topics would expedite both national level capacity building as well as will foster south-south exchange of learning.

A comprehensive and robust adaptation planning require application of different methodologies and tools. Current trend to apply different methodologies and tools is project based and often supported by external experts and institutes. Therefore, it does not build and leave much capacity at national level particularly institutional capacity. Capacity gaps are often evident in terms of (a) collection, analysis, interpretation, and utilization of climate data; (b) climate modelling; and (c) assessment of technologies. To address current technical capacity gaps on application of different methodologies and tools such as for assessing vulnerabilities of different sectors and at different levels, appraising adaptation options and prioritizations, etc. building capacity of national institutes to support adaptation planning is essential. While immediate actions are necessary to address the current gaps, development of climate change curriculum and integration into the education system will enable to produce future actors who will have adequate academic understanding as well as application of different methods and tools to analyse complex situations.

A mechanism to undertake tangible indicators based monitoring and reporting system to regularly undertake stocktaking exercise and reporting progress of adaptation is essential. It is also necessary to empower national institutes to plan and implement this monitoring and reporting mechanism. Without an indicator-based approach with a clear set baseline, monitoring and reporting progress of existing policies and programmes, and interventions would be difficult. Further, monitoring and reporting indicators need to be revisited time to time and adjust them considering expected changes in climate scenarios, population dynamics, socio-economic development, and national commitments to global climate change processes. In addition to monitoring and reporting progress of achievement, there is a need to record and monitor the outcome of the capacity building activities. It would also require benchmark indicators and monitor improvement of capacity against those indicators.

11.2.5 Improve Access and Paradigm Shift in Climate Finance

Almost all countries in Asia and the Pacific have access adaptation finance mostly through international accredited entities except couple of recent exceptions of accessing climate finance through direct access entities. Accessing climate finance directly through national entity is becoming a desired option for which greater efforts are needed to overcome existing barriers related to fiduciary management and compliance with social and environmental norms set by Adaptation Fund and Green Climate Fund (GCF).

While most of the countries in Asia and the Pacific region have identified their adaptation needs, access to adaptation funds is one of the key determinants of implementation of adaptation measures. Amount available also plays an important role on scale of adaptation intervention. Usually, availability and access to adaptation finance often refers multilateral funds such as Least Developed Country Fund (LDCF), Special Climate Change Fund (SCCF), and Green Climate Fund (GCF) established under the framework convention on climate change. One of the commonly identified challenges is in-country technical capacity to develop adaptation project as well as lack of vulnerability assessment at a geographical scale for which adaptation interventions are to be designed.

Considering scale of financial resources require for climate change adaptation, and its interlinkage with development, in addition to improve access climate finance from international sources it is important to explore opportunities to utilize national budget. Even a small share of the national fund increases the ownership of the national counterparts and will encourage mainstreaming climate responses into development and sector-based planning. One of the mechanisms, supporting country and community-based interventions on adaptation could be the establishment of the National Climate Trust Fund. Several countries in the region have established climate change fund. For examples, Bangladesh Climate Change Trust Fund as well as Climate Change Resilience Fund, and the People's Survival Fund in Philippines. These types of national funds provide more predictable financing and take proactive measures for adaptation.

Further, considering the scale of funding requirement, and most importantly strong linkage between climate change adaptation and development investment, there may be a need of redirect investments and fund flows from all sources to more sustainable climate-friendly interventions. There is need for scaling-up of international private and public finance and optimization of the funds allocation by spreading the risk across private and public investors.

11.2.6 Enhance Cooperation in Technology Transfer

Except a few countries in Asia and the Pacific region, Technology Needs Assessment (TNA) and development of Technology Action Plan (TAP) are available. However, a country needs to regularly update technology needs assessment and action plan to adjust them with social and economic development situation of the country as well as incorporate advancement of technologies and climate change knowledge and understanding. At present, most countries in the region are applying several tested technologies for adaptation to climate change, e.g. climate (drought, flood, salinity) resilient crop varieties, water conservation and saving techniques, soil and land management, etc. Low-tech and low-cost irrigation system for isolated communities in the Pacific as well as rural areas in other countries become a sustainable practice as farmers can easily set up and manage this technology. Establishment of supporting institution at the local level such as the *Kivori* Cooperative Society in Papua New Guinea to assist local communities appears beneficial for farmers in addressing challenges and to expand and reach new markets. However, many of them are at pilot scale and sporadic.

Use of information and communication technologies (ICTs) for different purposes such as in agriculture and disaster risk reduction are providing enormous benefits where it is being applied successfully. In Japan, mobile farming creates a community and a network linking farmers, farmer's organizations, buyers, local governments, and the scientific community. It offers flexibility and access to opportunities to coordinate with multiple stakeholders for innovative agricultural practices, sustainable land management, good e-governance, and financial stability. The emergence of new ICTs and their application in agriculture helps decision-makers to respond during emergency situations. Consistent with the goals of the society to become sustainable, ICT tools should be promoted in other countries especially in the community level.

Enhance cooperation among countries in the region on technology transfer, share knowledge, and experiences on climate adaptation technologies are very much in need, not only within the country and community but at the regional scale. Further, it requires support for replication as well as research to understand effectiveness of these technologies in long run as well as greater efforts to innovate and adopt new and advanced technologies.

Technology innovation and transfer often discussed as a need of developing countries to deal with adverse impacts of climate change. In technology innovation and transfer, role of government cannot be substituted by other approach and means. It is also important to note that transfer of the adaptive technology, for example, technologies developed for agriculture sector in Korea cannot be simply transferred to other countries. Therefore, successful innovation and transfer of technology needs to adopt a co-evolutionary learning approach based on the social and ecological systems already rooted in each society.

11.2.7 Enhance South-South Cooperation

Progress of adjustment of different elements to support adaptation actions varies by sub-region and countries which provides opportunity of south-south cooperation and learn from each other on the topics of common interests and highest concern such as agriculture and food security, water resources, and natural ecosystems. Share lessons on establishment of institutional structure and governance, good adaptation practices, and mainstreaming into development planning budgeting system would be worth pursuing. Ongoing networks and platforms including Asia Pacific Adaptation Forum under Asia Pacific Adaptation Network (APAN) provide excellent opportunity to the countries to learn from each other.

11.2.8 Utilize Benefits of Commonalities and Complementarities

In 2015, multilateral intergovernmental processes have delivered four major global agreements, i.e. Finance for Development in Addis, Sustainable Development Goals in New York, Paris Agreement (CoP21) in France, and Sendai Framework for Disaster Risk Reduction in Japan. Commonalities and complementarities do exist among these multilateral global agreements in terms of overarching objectives and timeframe to achieve several goals and targets under each agreement. Commonalities are also evident in terms of priority sectors, communities and peoples, and actors at different levels who will be engaged for implementation of more than one agreements. Nature of universality, scale of ambition, partnership for implementation, review, and progress monitoring are other key commonalities of these global agreements. Therefore, successful implementation of an agreement will reinforce in achieving objectives, goals, and targets of other agreements.

The commonly identified and prioritized sectors in SDG and Climate Changes are water, agriculture and food, human health, energy, transport, urban and housing. SDG aims to protect and restore water and water-related ecosystems including mountains, forests, wetlands, rivers, aquifers, and lakes, while climate change agreement and decisions aim to address seasonal variation and availability (abundance and lack), and quality of water including salinity intrusion. Types of measures and technologies identified include increasing water-use efficiency, integrated water resources management, desalinization of water, protecting coastal agriculture from salinity, and introduction of more salt-tolerant varieties. Introduction of these measures and technologies will reinforce both SDG and climate change in achieving its goals and targets related to poverty alleviation, hunger, availability of food, and maintaining good freshwater ecosystem and ecosystem services.

Sustainably manage and protect marine and coastal ecosystems to avoid significant adverse impacts and take action for restoration in order to achieve healthy and productive oceans as well as minimize the impacts of ocean acidification are also common. Protection, restoration, and promotion of sustainable use of terrestrial ecosystems, dealing with land degradation and loss of biodiversity, conservation, restoration, and sustainable use of terrestrial and inland freshwater ecosystems and their services are in line with obligations under international agreements. Promotion of sustainable management of forests, dealing with deforestation and restoration of degraded forests and increase afforestation are common areas as well.

Building resilient infrastructure, make cities and human settlements resilient and sustainable through upgrading infrastructure and retrofit industries, increase resource-use efficiency and greater adoption of clean and environmentally sound technologies and industrial processes are common areas both in SDG and Climate Change. Goal 13 of SGD is very specific to climate change and agreed to take urgent action to combat climate change and its impacts, strengthen resilience and adaptive capacity to climate-related hazards and natural disasters, and integrate climate change measures into national policies, strategies, and planning which have mirrored in Paris Agreement on Climate Change.

Types of review, monitoring progress and reporting outcomes of implementation of actions, and legal nature and level of compliance of these multilateral global agreements varies ranging from voluntary to legally binding nature but the primary responsibility lies on the government for review and monitoring progress, and reporting results/achievement at national, regional, and global levels. It is evident that among these agreements, legal nature, and compliance mechanisms of climate change agreement is more rigorous than other agreements.

Countries are taking steps in domesticating these global outcomes into their national development sphere including integration into policies and plans, identification of capacity needs to implement actions and report progress, and access to technology and finance for delivering these global outcomes. Considering complementary nature, an overarching institutional coordination mechanism will help effective implementation of different measures, ensure synergies and avoid duplications of efforts. Moreover, the overarching objectives of climate change convention, and goals and targets under the sustainable development agenda are similar in terms of the sectors as well as the technological interventions. Therefore, institutionalization of climate change-related targets and sustainable development-related targets is very important.

11.3 Conclusion

Paris Agreement on climate change presents a new hope with political commitment and aspiration to achieve temperature goals, i.e. increase temperature below 2 °C and aspire for 1.5 °C through a bottom-up approach called "Nationally Determined Contribution (NDC)". This agreement is considered as a breakthrough in the climate change negotiations, which has been accepted by 195 countries with number of global goals and targets; means of implementation; and adequate review mechanisms and tools to monitor progress both at international and national levels. The

Agreement has also been ratified by 176 countries, as of 27 May 2018 (UNFCCC 2018). Parties have also agreed to submit and update periodically an adaptation communication, which may include country's priorities, implementation, and support required for adaptation plans and actions. Global Stocktake under Paris Agreement will also consider information in the adaptation communication for enhancing the implementation of adaptation action.

Considering varied governance structure among countries in Asia and the Pacific region, institutional coordination mechanism focusing on development priorities and interventions that benefit both climate change and SDGs, clarity on roles and responsibilities of actors for implementation of actions at national level and reporting to international bodies is an urgent necessity. More stable mechanism by institutionalizing implementation aspects of achieving the goals and targets at national levels, brining international support in more coordinated manner, and reporting progress at international level would require. It will also help and support implementation and to avoid duplication of efforts and efficacy of resources use. The role of private sector and their engagement in financing for technologies and technologies deployment are critical for which action requires by international and national actors to create enabling environment.

Being the host of large world population with significant diversity in society and culture, economy and institutions, resource endowment and degradation, vulnerability and adaptive capacity to adverse impacts climate change, etc. adaptation journey ahead of the region is challenging and long as well. Diversity in many aspects and continued need of development efforts present opportunity for a country to integrate climate change into development planning and budgeting system. It is high time for the country to start viewing adaptation to climate change within the overall development framework.

References

IPCC (2014) Climate change 2014: impacts, adaptation, and vulnerability
UNFCCC (2018) https://unfccc.int/process/the-paris-agreement/status-of-ratification

Index

M. Alam et al. (eds.), *Status of Climate Change Adaptation in Asia and the Pacific*, Springer Climate, https://doi.org/10.1007/978-3-319-99347-8

The manufacturer's authorised representative in the EU is Springer
Nature Customer Service Centre GmbH, Europaplatz 3, 69115 Heidelberg,
Germany. If you have any concerns regarding our products, please
contact ProductSafety@springernature.com

Printed and bound by CPI Group (UK) Ltd, Croydon, CR0 4YY
01/10/2026
02225975-0010